AF361775

Green Concrete for a Better Sustainable Environment

Green Concrete for a Better Sustainable Environment

Special Issue Editor

Patrick Tang

MDPI • Basel • Beijing • Wuhan • Barcelona • Belgrade • Manchester • Tokyo • Cluj • Tianjin

Special Issue Editor
Patrick Tang
The University of Newcastle
Australia

Editorial Office
MDPI
St. Alban-Anlage 66
4052 Basel, Switzerland

This is a reprint of articles from the Special Issue published online in the open access journal *Applied Sciences* (ISSN 2076-3417) (available at: https://www.mdpi.com/journal/applsci/special_issues/Green_Concrete_Sustainable_Environment).

For citation purposes, cite each article independently as indicated on the article page online and as indicated below:

LastName, A.A.; LastName, B.B.; LastName, C.C. Article Title. *Journal Name* **Year**, *Article Number*, Page Range.

ISBN 978-3-03936-234-9 (Hbk)
ISBN 978-3-03936-235-6 (PDF)

© 2020 by the authors. Articles in this book are Open Access and distributed under the Creative Commons Attribution (CC BY) license, which allows users to download, copy and build upon published articles, as long as the author and publisher are properly credited, which ensures maximum dissemination and a wider impact of our publications.

The book as a whole is distributed by MDPI under the terms and conditions of the Creative Commons license CC BY-NC-ND.

Contents

About the Special Issue Editor

Patrick Tang is an Associate Professor at the School of Architecture and Built Environment (SABE) of the University of Newcastle. Associate Professor Tang specializes in concrete technology and design. He has extensive materials science research experience and his work includes in-depth analyses from the microstructure to the macro-engineering behavior of materials, products and structures related to sustainable cement-based composites. Tang is currently the SABE Research Director, responsible for developing and implementing strategies to support research performance and engagement across the School.

Editorial

Special Issue on Green Concrete for a Better Sustainable Environment

Waiching Tang

School of Architecture and Built Environment, University of Newcastle, Callaghan, NSW 2308, Australia; patrick.tang@newcastle.edu.au; Tel.: +61-249-217-246

Received: 30 March 2020; Accepted: 5 April 2020; Published: 9 April 2020

1. Introduction

Green concrete is defined as concrete that uses waste material as at least one of its components, or has a production process that does not lead to environmental destruction, or has a high performance and life cycle sustainability. At present, natural resources are running out. Using industrial and construction waste as raw materials for the production of cement and concrete can be regarded as a valuable resource for civil infrastructure construction. Green concrete will not only contribute to a circular economy but can also help to reduce the amount of embodied energy and CO_2 emissions associated with cement manufacturing, aggregate quarrying and to mitigate the environmental threats associated with industrial waste materials. This Special Issue aims to cover recent advances in the development of green concrete solutions and deliberate on what can best be done to leverage the opportunities. The papers published in this special issue cover theoretical, experimental, applied and modelling studies that research on the materials, products and the structures related to sustainable cement-based composites. The papers are categorized into several representative themes and the main contents of each paper are briefly summarized.

2. Credible Long-Term Performance and Durability

Impact of Temperature Changes and Freeze—Thaw Cycles on the Behaviour of Asphalt Concrete Submerged in Water with Sodium Chloride [1]

This paper studies the mechanical behaviour of an asphalt concrete when it is subjected to temperature changes and freeze–thaw cycles. The results show that, although the temperature changes have a negative effect on the mechanical properties, salt water protects the aggregate-binder adhesive, maintains the mechanical strength, increases the number of load cycles for any strain range and reduces the time that the mixture is in contact with frozen water.

MSWI Bottom Ash Application to Resist Sulfate Attack on Concrete [2]

This research provides a strategy for partially replacing cement with municipal solid waste incineration (MSWI) bottom ash (BA) to improve the performance of concrete against sulphate attack. The results show that the replacement of cement with BA can improve the durability of concrete and actualize the utilization of MSWI residues as a resource.

Thermal and Mechanical Properties of Cement Mortar Composite Containing Recycled Expanded Glass Aggregate and Nano Titanium Dioxide [3]

This paper aims to investigate the effects of recycled expanded glass aggregates (EGA) on fresh and hardened properties and the thermal insulating performance of cement mortar. The results demonstrate that EGA-mortar can be integrated into the building envelop or non-load-bearing-elements, such as wall partition as a thermal resistance, to reduce the long-term energy consumption in residential buildings.

3. Proven Structural Reliability

Experimental Study on the Seismic Performance of Recycled Concrete Hollow Block Masonry Walls [4]

This paper aims to manufacture a new recycled concrete hollow block (RCHB) which can be used for masonry structure with seismic requirements. The influences of the aspect ratio, vertical axial stress and the different materials used for structural columns on the seismic performance of RCHB masonry walls were studied. The research confirms that RCHB masonry walls could meet the seismic requirements through thoughtful design.

Experimental Study on Seismic Behavior of Steel Frames with Infilled Recycled Aggregate Concrete Shear Walls [5]

Experiments were performed on four specimens of steel frames with infilled recycled aggregate concrete shear walls (SFIRACSWs), one specimen of infilled ordinary concrete wall as well as one pure-steel frame and were conducted under horizontal low cyclic loading. The main seismic performance indexes of SFIRACSWs are evaluated comprehensively, providing a theoretical basis for popularizing and applying SFIRACSWs in practical engineering.

4. Reliable Numerical and Modelling Studies

An Experimental Strain-Based Study on the Working State of Husk Mortar Wallboards with Openings [6]

This paper tested full-scale HMES wallboards with different openings and obtained the strains at points distributed on the wallboard sides. An empirical relationship among the experimental failure loads of the wallboards was derived based on the failure characteristics of the wallboards, which could provide a reference to the accurate prediction of the wallboards' load-bearing capacity.

Numerical Study of Bond Slip between Section Steel and Recycled Aggregate Concrete with Full Replacement Ratio [7]

In this paper, the bond deterioration mechanism of recycled aggregate concrete (RAC) with a full replacement ratio was studied through experimental and numerical simulations. The analysis of the results showed that the developed model is capable of representing the characteristic bond strength value between section steel and RAC with sufficient accuracy.

Experimental Study on a Prediction Model of the Shrinkage and Creep of Recycled Aggregate Concrete [8]

In this paper, 180-day shrinkage and creep tests of recycled aggregate concrete with different water–cement ratios were designed in order to analyse the effect of the substitution rate and water–cement ratio on the shrinkage and creep properties. Moreover, the shrinkage and creep models of the recycled aggregate concrete were established.

Acknowledgments: This issue would not have been possible without the help of a variety of talented authors, professional reviewers, and the dedicated editorial team of Applied Sciences. Thank you to all the authors and reviewers for this opportunity. Finally, thanks to the Applied Sciences editorial team.

Conflicts of Interest: The author declares no conflict of interest.

References

1. Vega-Zamanillo, Á.; Juli-Gándara, L.; Calzada-Pérez, M.; Teijón-López-Zuazo, E. Impact of Temperature Changes and Freeze—Thaw Cycles on the Behaviour of Asphalt Concrete Submerged in Water with Sodium Chloride. *Appl. Sci.* **2020**, *10*, 1241. [CrossRef]
2. Cheng, Y.; Dong, Y.; Diao, J.; Zhang, G.; Chen, C.; Wu, D. MSWI Bottom Ash Application to Resist Sulfate Attack on Concrete. *Appl. Sci.* **2019**, *9*, 5091. [CrossRef]

3. Yousefi, A.; Tang, W.; Khavarian, M.; Fang, C.; Wang, S. Thermal and Mechanical Properties of Cement Mortar Composite Containing Recycled Expanded Glass Aggregate and Nano Titanium Dioxide. *Appl. Sci.* **2020**, *10*, 2246. [CrossRef]
4. Liu, C.; Nong, X.; Zhang, F.; Quan, Z.; Bai, G. Experimental Study on the Seismic Performance of Recycled Concrete Hollow Block Masonry Walls. *Appl. Sci.* **2019**, *9*, 4336. [CrossRef]
5. Sun, L.; Guo, H.; Liu, Y. Experimental Study on Seismic Behavior of Steel Frames with Infilled Recycled Aggregate Concrete Shear Walls. *Appl. Sci.* **2019**, *9*, 4723. [CrossRef]
6. Cai, X.; Sun, S.; Zhou, G. An Experimental Strain-Based Study on the Working State of Husk Mortar Wallboards with Openings. *Appl. Sci.* **2020**, *10*, 710. [CrossRef]
7. Liu, C.; Xing, L.; Liu, H.; Quan, Z.; Fu, G.; Wu, J.; Lv, Z.; Zhu, C. Numerical Study of Bond Slip between Section Steel and Recycled Aggregate Concrete with Full Replacement Ratio. *Appl. Sci.* **2020**, *10*, 887. [CrossRef]
8. Lv, Z.; Liu, C.; Zhu, C.; Bai, G.; Qi, H. Experimental Study on a Prediction Model of the Shrinkage and Creep of Recycled Aggregate Concrete. *Appl. Sci.* **2019**, *9*, 4322. [CrossRef]

 © 2020 by the author. Licensee MDPI, Basel, Switzerland. This article is an open access article distributed under the terms and conditions of the Creative Commons Attribution (CC BY) license (http://creativecommons.org/licenses/by/4.0/).

 *applied sciences*

Article

Impact of Temperature Changes and Freeze—Thaw Cycles on the Behaviour of Asphalt Concrete Submerged in Water with Sodium Chloride

Ángel Vega-Zamanillo [1], Luis Juli-Gándara [1,*], Miguel Ángel Calzada-Pérez [1] and Evelio Teijón-López-Zuazo [2]

[1] GCS Research Group, Civil Engineering School, Universidad de Cantabria, 39005 Santander, Spain; vegaa@unican.es (Á.V.-Z.); calzadam@unican.es (M.Á.C.-P.)

[2] Construction and Agronomy Department, University of Salamanca, 49022 Zamora, Spain; eteijon@usal.es

* Correspondence: ljg59@alumnos.unican.es or luis_juli@hotmail.com

Received: 21 January 2020; Accepted: 9 February 2020; Published: 12 February 2020

Abstract: One of the main applications of salt in civil engineering is its use as a de-icing agent on roads in cold areas. The purpose of this research is to find out the mechanical behaviour of an asphalt concrete when it is subjected to temperature changes and freeze–thaw cycles. These temperature interactions have been carried out for dry specimens, specimens submerged in distilled water and specimens submerged in salt water (5% of sodium chloride, NaCl). An AC16 Surf D bituminous mixture was evaluated under three types of temperature interaction: three reference series remained at a controlled temperature of 20 °C, another three series were subjected to five freeze–thaw cycles and the last three series have been subjected to one year outside in Santander (Spain). The mechanical behaviour of the mixture was determined by Indirect Tensile Strength Test (ITS), Water Sensitivity Test (ITSR) and Wheel Tracking Test, Dynamic Modulus Test and Fatigue Tests. The results of the tests show that, although the temperature changes have a negative effect on the mechanical properties, salt water protects the aggregate-binder adhesive, maintains the mechanical strength, increases the number of load cycles for any strain range and reduces the time that the mixture is in contact with frozen water.

Keywords: salt; NaCl; asphalt concrete; freeze–thaw cycles; winter road

---

Highlights: Specimens of hot mix asphalt were evaluated under different temperature changes, including freeze–thaw cycles. Temperature changes have a harmful effect on the behaviour of the mixture, but the amount of time that the mixture is submerged in contact with salt water is the main mechanism of damage. Salt water protects the aggregate-binder adhesive, maintains the mechanical strength, increases the number of load cycles for any strain range and reduces the time that the mixture is in contact with frozen water.

1. Introduction

Currently, the use of sodium chloride (NaCl) is widespread in a large number of countries, as a de-icing agent on winter roads. The impact that salt has on the environment has been extensively studied; the melting salt leads to the mobilization of heavy metals such as lead, cadmium, copper and zinc [1–4], along with increased chlorides (Cl^-) [5,6]. However, salt is still used, due to its low cost, versatility and physical properties [7,8]. These physical properties make their use propitious in temperatures up to −21 °C, but as García [7] indicated, from −5 °C, in order to be more effective, they can be used in combination with calcium chloride ($CaCl_2$).

Shi et al. [9] noted that the value of negative impacts on the environment and on vehicles must be taken into account when calculating the costs of de-icing agents, including NaCl. In this trend of reducing costs and negative impacts, Klein-Paste et al. [10] indicated that the amount of NaCl that is used in practice could be reduced by 40%, to prevent the pavement from being slippery, because the salt spread on the roads causes a weakening of the ice. Another aspect studied along the line of reducing the negative impact of NaCl is the creation of tools to decide more efficiently when salt should be applied to roads [11,12].

There are also studies on the behaviour of bituminous mixtures in contact with the different de-icing agents. Wang et al. [13] noted that, compared to other anti-icing agents, such as sand and quartz dust, salt does not produce a polished surface. Hassan et al. [14] indicated that the value of Indirect Tensile Strength (ITS) for asphalt mixtures exposed to salt after 25 freeze–thaw cycles is similar to other de-icing agents, such as potassium acetate, sodium formate or urea. However, after 50 cycles, the ITS result is even better for the mixtures exposed to salt than those submerged in distilled water.

Feng et al. [15] studied the impact of salt on the performance of bituminous mixtures when it remains submerged under seawater subjected to freeze–thaw cycles. They simulate this fact by adding salt to bitumen and subjecting various types of mixtures (AM-16, OGFC-19 and AC-16) to freeze–thaw cycles; their results show a decrease in the Water Sensitivity Test (ITSR), with results being lower in the case of the AC mixture.

Juli-Gándara et al. [16] investigated how NaCl influences the mechanical properties of three types of asphalt mixtures: a hot mix asphalt with conventional bitumen; and two porous mixtures, one manufactured with a conventional binder and another with a modified binder. The effect of the salt is analysed by three different processes: immersing the specimens in salt water; adding salt as aggregate into the mixture; and submerging the aggregate in water with a certain concentration of salt, drying it and then making the mixture with it. The results show that the hot mix asphalt is scarcely affected when it is submerged in salt water.

The effects of freeze–thaw cycles in the bituminous mixture are widely studied. Goh and You [17] used an image-processing technique to indicate that the removal of fine and coarse aggregates on the surface increases when the asphalt mixture undergoes more freeze–thaw cycles. Tarefder et al. [18] tested an AC mixture to ITS and Fatigue Test after 5, 10, 15 and 20 freeze–thaw cycles, and the results showed a reduction of 30.7% in the fatigue life and a small amount of reduction of ITSR after five cycles. Özgan and Serin [19] indicated through the void ratio (Vh), the void ratio filled with asphalt (Vf) and the void ratio inside mineral aggregate (VMA) parameters; the ultrasonic velocity test; and the Marshall Stability (MS) that the effect of the freeze–thaw cycles on the asphalt concrete is highly important, especially for the hot mix asphalt design. Islam and Tarefder [20], who investigated the stiffness and tensile strength degradation behaviour of asphalt concrete on long-term freeze–thaw samples in the laboratory, indicated that the flexural stiffness decreases with the number of freeze–thaw cycles, whereas the ITS does not change significantly with the number of freeze–thaw cycles. Teltayev et al. [21] investigated, in laboratory conditions, the effect of cyclic freezing and thawing on the characteristics of the neat bitumen and bitumens modified with different polymers, as well as stone mastic asphalt concretes. Their results of the SMA-20 Bit 100/130 show a 78% decrease in the value of ITS and a 270% increase in the Rut Depth (RD) after 50 freeze–thaw cycles.

Until now, no research has covered the results of mechanical behaviour such as mechanical strength, permanent deformations, dynamic modulus and fatigue life, when an asphalt concrete mixture is subjected to temperature changes, freeze–thaw cycles and the impact of salt. The purpose of this research is to fill this gap. The temperature interactions have been carried out for dry specimens, specimens submerged in distilled water and specimens submerged in salt water (NaCl).

2. Methodology

2.1. Materials

2.1.1. Aggregate

The aggregate used in this research is an ophite with the following properties (Table 1):

Table 1. Properties of aggregate.

Aggregate	Los Angeles Abrasion Test (%) (UNE-EN 1097-2:2010) [22]	Water Absorption (%) (UNE-EN 1097-6:2014) [23]	Flakiness Index (%) (UNE-EN 933-3:2012) [24]
Ophite	16.0	1.0	9.0

The filler (mineral powder) is limestone, with a specific weight of 2.753 g/cm^3.

2.1.2. Binder

A conventional bitumen B 50/70, which is frequently used in Spain, was used. Table 2 shows the principal properties of the bitumen:

Table 2. Properties of bitumen.

Bitumen	Penetration (25 °C; 100 g, 5 s) (0.1 mm) (UNE-EN 1426:2015) [25]	Softening Point (°C) (UNE-EN 1427:2015) [26]	Frass Breaking Point (°C) (UNE-EN 12593:2015) [27]
B 50/70	65.0	47.2	−9

2.1.3. Mixture

An AC16 Surf B50/70 D mixture, which is frequently used on surface pavements in Spain, was studied. The composition is shown in Table 3. The density of the mixture, according to the Marshall Test (UNE-EN 12697-34:2013) [28], is 2.458 g/cm^3.

Table 3. Composition of bituminous mixture.

	Passing Rate (%)								Bitumen Content
Sieve size (mm)	22	16	8	4	2	0.5	0.25	0.063	*s/m* (%)
AC16 Surf D	100.0	95.0	71.5	51.5	38.5	21.5	15.5	6.0	5.0

The manufacture of the AC16 Surf B 50/70 D is done at the SENOR S.A. asphalt plant.

2.1.4. Salt

The salt used for this work is NaCl. Table 4 indicates the particle size of salt. The specific weight of the salt used is 2.165 g/cm^3.

Table 4. Particle size of salt.

	Passing Rate (%)							
Sieve Size (mm)	22	16	8	4	2	0.5	0.25	0.063
Salt	100.0	100.0	100.0	100.0	100.0	74.9	39.9	6.6

2.2. Experimental Plan

Nine different series have been evaluated. All the batches of bituminous mixture were manufactured at the same time. After this, the specimens were subjected to individual analysis (Table 5):

- Test Series A: remains at a constant temperature of 20 °C;
- Test Series B: subjected to five freeze–thaw cycles;
- Test Series C, subjected to one year outside storage;

whereby

- (A0) is dry specimens,

- (A1) is submerged in distilled water,
- (A2) is submerged in salt water,

in each case.

Table 5. Series description.

		Water Interaction		
		Dry	**Submerged in Distilled Water**	**Submerged in Salt Water**
Temperature interaction	Constant temperature at 20 °C	A0	A1	A2
	Five freeze–thaw cycles	B0	B1	B2
	One year outside storage	C0	C1	C2

The amount of salt by water weight in A2, B2 and C2 series is five percent (5.0%). For submerged series, the temperature and time that the specimens are submerged change according to the test requirements.

2.2.1. Temperature Interaction

Three different temperature interactions were analysed. The A-Series are for reference; the temperature remains constant at 20 °C.

Freeze–Thaw Cycles

The B-Series were subjected to five freeze–thaw cycles between 38 and −33 °C. These temperatures have been chosen because 38 °C is approximately ten degrees higher than the summer air temperature average in Santander (Spain), and −33 °C is ten degrees lower than the freezing point of water with 5.0% salt content.

The temperature was measured by introducing a thermo-sensor inside the specimens at the time of manufacturing the mixture (Figure 1). With these thermo-sensors, it is possible to determine the internal temperature of the specimens (logging interval one minute). Knowing this, the time of each freeze–thaw cycle is established so that both phases, hot and cold, have enough time to reach the defined temperatures. Each freeze–thaw cycle requires 48 h to be completed, 24 h for the hot time period and 24 h for the cold time period (Figure 2).

Figure 1. Thermo-sensor inside the specimens.

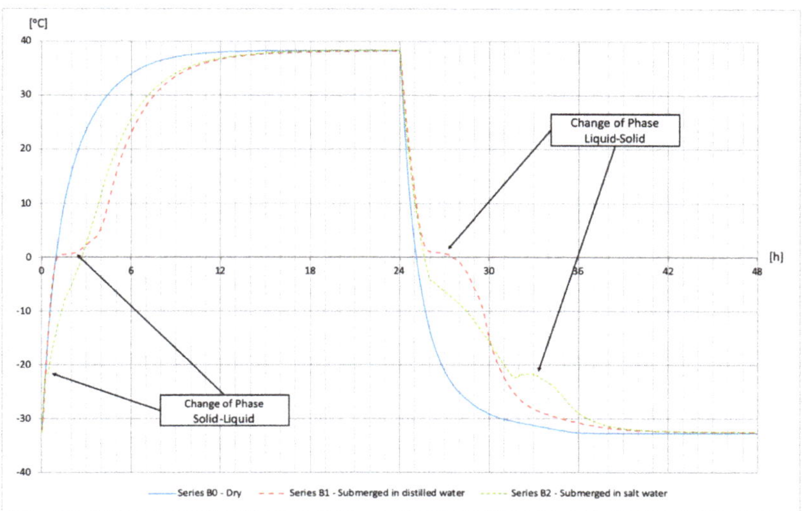

Figure 2. Freeze–thaw cycle.

The time that each submerged series remained in frozen water was defined by the steps of solid–liquid and liquid–solid changes. In the case of the B1 series, the time submerged in frozen water is 22 h, and in the B2 series, it is approximately 15 h.

Outside Storage

The outside storage series, the C-Series, remained outside in Santander (Spain) for one year, from 20 December 2016 to 20 December 2017, and was protected from precipitation. The maximum and minimum daily temperatures during this period were collected by the Spanish Meteorological Agency AEMET (Figure 3).

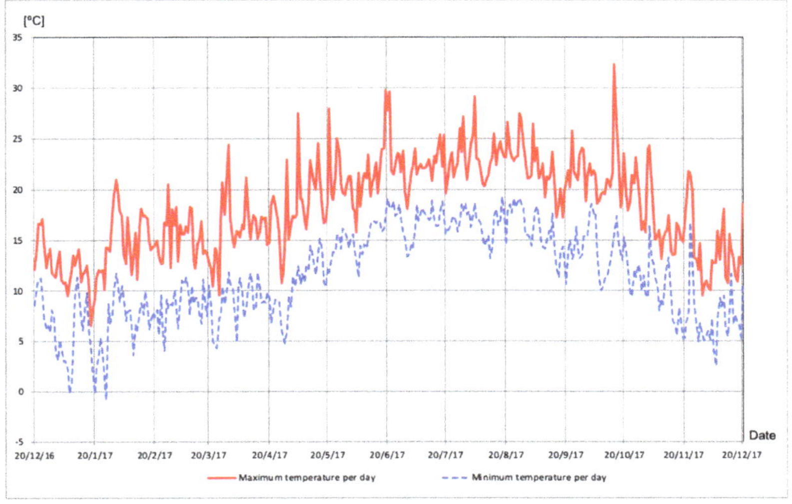

Figure 3. Maximum and minimum temperature per day in Santander for the year 2017.

2.3. Tests Programme

2.3.1. Indirect Tensile Strength (ITS) and Water Sensitivity Test (ITSR)

These tests have been done to determine the mechanical strength and how the aggregate-binder adhesive is influenced by the action of water. As established by Spanish Standard PG-3 [29], the test procedure was the "Method A" of the UNE-EN 12697-12: 2009 standard [30], with a compaction of the specimens by impact (UNE-EN 12697-30: 2013) [31], with 50 blows per side. Four dry specimens and another four wet specimens for each series were evaluated.

The dimensions of the specimens are 63.5 mm in height and 101.6 mm in diameter. This test is done for all specimens.

2.3.2. Wheel Tracking Test

This test has been carried out to determine the susceptibility of the specimen to deformation when a moving vertical load is applied. The test is done in accordance with the standard UNE-EN 12697-22: 2008 + A1 "Procedure B in air" [32], compacted by roller compactor (UNE-EN 12697-33: 2006 + A1) [33]. The limits established by PG-3 depending on the heavy traffic categories are 0.07 and 0.10 mm/10^3 cycles of Wheel Track Slope (WTS).

The specimens are 410 mm long by 260 mm wide by 50 mm high. The Wheel Tracking Test is only done for the A-Series and C-Series. For these series, two specimens of each one were tested.

2.3.3. Dynamic Modulus Test

Since bituminous mixtures are a viscous material, this test was done to provide insight into the viscosity properties of the mixtures. The test was carried out as established by the standard UNE-EN 12697-26: 2012 "Annex B" four-point bending test [34]. Six specimens have been evaluated for each series, except for the C-Series, where eight specimens have been tested due to the long period that these specimens have been outdoors.

The specimens are 410 mm long by 60 mm wide by 60 mm high. The Dynamic Modulus Test was done for all specimens.

2.3.4. Fatigue Test

This test provides an estimate of durability of the bituminous mixture by load–unload cycles. The test procedure has been the UNE-EN 12697-24: 2013 "Annex D" four-point bending test [35]. The same number of specimens, as in the case of the Dynamic Modulus Test, was evaluated.

The dimensions of the specimens are the same as the ones of the Dynamic Modulus Test. The Fatigue Test is done for all specimens.

3. Results and Discussion

3.1. Indirect Tensile Strength (ITS) and Water Sensitivity Test (ITSR)

Table 6 shows the results of Indirect Tensile Strength and Water Sensitivity Test for all the series. The same as Hassan et al. [14] indicated, the values of ITS are similar or even better for mixtures exposed to salt than those submerged in distilled water.

The results obtained in the A-Series for ITS and ITSR are the regular ones for a hot mix asphalt. There are no noticeable differences between the series submerged in distilled water (A1) and the series submerged in salt water (A2). This result was expected due to the fact that the A-Series was not submitted to temperature changes.

In the series subjected to five freeze–thaw cycles (B-Series), although the result of ITS for dry specimens (B0) shows no significant difference when compared to the dry reference specimens (A0), that difference exists for the wet series. This variation is lower between specimens submerged in salt water (B2 and A2) and greater between specimens submerged in distilled water (B1 and A1). These results show that the greatest damage to the hot mix asphalt is not the temperature changes but the amount of time that the mixture is submerged in frozen water. The B2 series obtained a value of ITRS, which is close to the limit established by PG-3 (Spanish Regulation) for surface layers (85%).

Table 6. Indirect tensile strength and water sensitivity.

	Series	AC16 Surf B 50/70 D		
		Maximum Load (kN)		ITSR (%)
Constant temperature at 20 °C	A0	2340	–	–
	A1	–	2075	89
	A2	–	2053	88
Five Freeze–Thaw Cycles	B0	2385	–	–
	B1	–	1540	63
	B2	–	1975	83
One year outside storage	C0	2189	–	–
	C1	–	1418	65
	C2	–	1690	77

For the series subjected to a year outdoors (C-Series), the results of ITS for dry and submerged specimens are lower than the ones of the reference series (A-Series). However, in the case of submerged specimens (C1 and C2), especially for the series submerged in distilled water, the values of the ITS test are lower than in dry specimens (C0). This result is due to the fact that remaining submerged for a year is a more damaging process for the mixture than remaining dry for the same period.

3.2. Wheel Tracking Test

The values for the Wheel Track Slope and the Ruth Depth at 10,000 cycles (RD) for all the series are shown in Tables 7 and 8. The results obtained from these tests comply with the Spanish regulations for most climate zones and traffic.

Table 7. Wheel track slope.

	Series	AC16 Surf B 50/70 D
		WTS (mm/10^3 Cycles)
Constant temperature at 20 °C	A0	0.08
	A1	0.07
	A2	0.05
One year outside storage	C0	0.06
	C1	0.04
	C2	0.08

Although the results for all series are similar, the series that have remained a year outdoors have obtained WTS and RD values lower than the reference series, except for the series whose specimens were submerged in salt water (C2). These low values (C0 and C1) can be due to the fact that, after one year in outside storage, the binder has become stiffer.

Table 8. Rut depth.

Series	AC16 Surf B 50/70 D	
		RD (mm)
Constant temperature at 20 °C	A0	3.2
	A1	3.6
	A2	2.4
One year outside storage	C0	2.6
	C1	2.1
	C2	3.0

3.3. Dynamic Modulus Test

As Table 9A,B indicate, the phase angles have no significant variation between the different series. Only in the case of those series which have been subjected to freeze–thaw cycles, the variation is slightly greater. This fact may be due to the larger temperature differences between the B-Series and the other series.

Table 9. Dynamic modulus test.

(A)

		AC16 Surf B 50/70 D								
		Frequency (Hz)								
	Series	0.1		0.2		0.5		1.0		2.0	
		E (GPa)	Phase Angle (°)	E (GPa)	Phase Angle (°)	E (GPa)	Phase Angle(°)	E (GPa)	Phase Angle (°)	E (GPa)	Phase Angle (°)
Constant temperature at 20 °C	A0	2.4	32.8	2.9	30.5	3.7	27.0	4.4	24.9	5.2	22.6
	A1	2.2	33.8	2.7	31.2	3.5	27.5	4.1	25.1	4.9	23.2
	A2	2.1	34.7	2.6	32.1	3.5	28.4	4.2	26.1	5.0	24.0
Five Freeze–Thaw Cycles	B0	2.2	35.9	2.7	33.2	3.5	30.0	4.3	27.4	5.1	25.2
	B1	1.7	34.8	2.2	32.1	2.9	28.9	3.5	26.8	4.1	24.6
	B2	2.0	33.0	2.5	30.5	3.3	27.1	4.0	25.0	4.7	23.0
One year outside storage	C0	2.4	31.3	2.9	28.9	3.7	26.0	4.4	23.9	5.2	21.9
	C1	2.2	33.1	2.6	30.6	3.4	27.5	4.1	25.4	4.8	23.3
	C2	2.2	34.0	2.8	31.2	3.6	28.0	4.3	25.8	5.1	23.6

(B)

		AC16 Surf B 50/70 D								
		Frequency (Hz)								
	Series	5.0		8.0		10.0		20.0		30.0	
		E (GPa)	Phase Angle (°)	E (GPa)	Phase Angle (°)	E (GPa)	Phase Angle (°)	E (GPa)	Phase Angle (°)	E (GPa)	Phase Angle (°)
Constant temperature at 20 °C	A0	6.2	20.3	6.9	19.0	7.1	18.5	8.1	17.0	8.9	16.5
	A1	6.0	20.7	6.6	19.2	6.9	18.6	7.9	17.6	8.7	16.6
	A2	6.1	21.3	6.7	20.0	7.0	19.3	8.0	17.8	8.9	17.6
Five Freeze–Thaw Cycles	B0	6.3	22.4	7.0	21.1	7.3	20.5	8.5	18.8	9.2	17.8
	B1	5.1	21.9	5.7	20.6	5.9	19.9	6.9	18.6	7.5	17.7
	B2	5.8	20.3	6.4	19.2	6.6	18.6	7.8	17.7	8.5	16.6
One year outside storage	C0	6.4	19.6	6.9	18.4	7.2	17.6	8.5	16.8	9.2	15.9
	C1	5.9	20.8	6.5	19.5	6.9	19.1	7.9	18.6	8.4	16.2
	C2	6.2	21.1	6.9	19.7	7.2	19.1	8.3	17.9	8.9	16.8

In the A-Series, very similar values were obtained for all the specimens. However, there exists a trend in which the dry series has higher modulus values and lower phase angles, due to the fact that it is more elastic than the two others.

However, for the series subjected to freeze–thaw cycles, a clear tendency appears; the dry specimens (B0) have greater modulus and phase angles for all of the test frequencies. In the case of the submerged series, B2 is more elastic than B1.

As in the case of the B-Series, for the C-Series, C0 has greater values of modulus, followed by C2 and finally by C1. However, in the case of the phase angles, the trend is different; the greatest angle is C2 and the lowest is C0. For the C-Series, the values of the modulus are greater, and the values of phase angles are lower compared with the A-Series and B-Series, and this indicates that the specimens have suffered a stiffening process.

In general, for all of the series and frequencies tested, the modulus values are greater for the dry series. This was expected due to the fact that submerging specimens in water is more harmful for the mixture than keeping them dry during any temperature interaction. Likewise, for all of the temperature interactions, there exists a trend which indicates that the series that are submerged in salt water have greater values of modulus than their corresponding pair submerged in distilled water, even more for the series subjected to freeze–thaw cycles. This trend supports the idea that the largest damage to the mixture that is submerged in water is the amount of time that it remains in frozen water, and the specimens that are submerged in salt water remain in frozen water for a shorter period of time.

3.4. Fatigue Test

The fatigue lines (Strain—Number of Load Cycles) obtained for all of the series are very similar, reaching notably high R^2 values; although, in the case of the C-Series, it is slightly lower. This may be due to the fact that the temperature interaction process is longer than in the A-Series and B-Series.

In the A-Series, for low values of strain, the submerged series, especially A1, obtain more load cycles than the dry specimens (Figure 4).

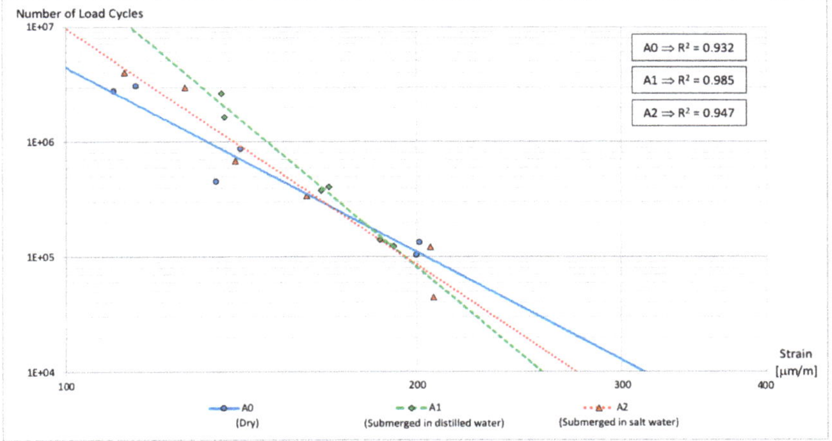

Figure 4. Fatigue test: specimens at constant temperature of 20 °C.

In the case of the series subjected to five freeze–thaw cycles (B-Series), the one that is submerged in salt water (B2) obtains more load cycles for any strain range than the other two series (Figure 5). This fact may be due to two main reasons. The first is that salt water offers a cushion effect on the temperature variations due to its thermal conductivity, which is smaller than that of distilled water [36]. The second is that the specimens of this series are in contact with frozen water for less time, which, as corroborated by the rest of the tests, is one of the most harmful effects for the mixture.

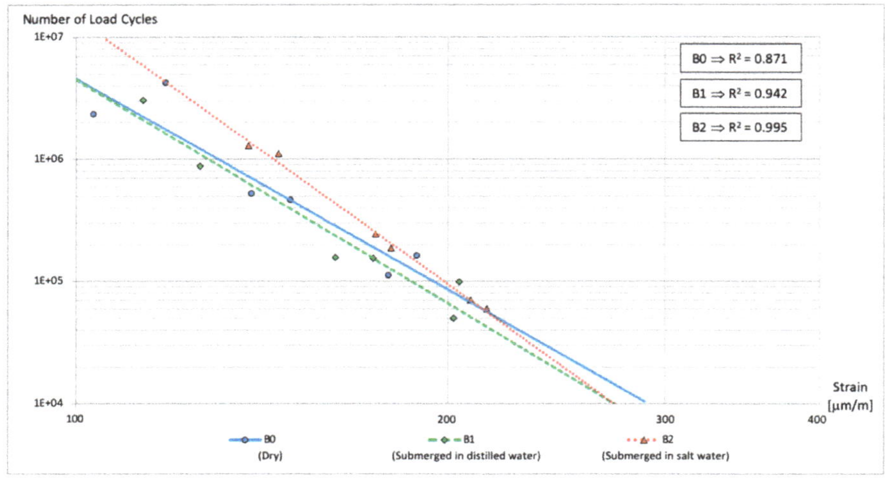

Figure 5. Fatigue test: specimens subjected to five freeze–thaw cycles.

The series that have been subjected to a year outdoors (C) obtain similar values for all specimens, being for C1 the number of load cycles slightly lower for high strain values (Figure 6).

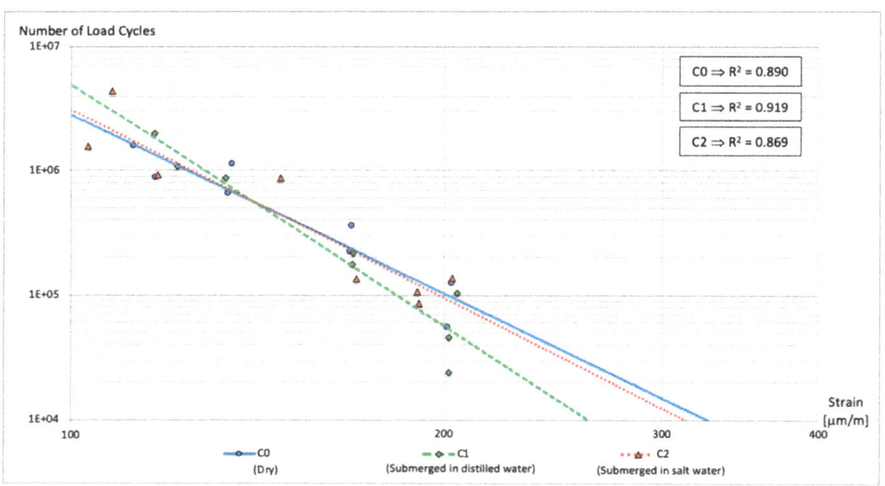

Figure 6. Fatigue test: specimens subjected to one of year outside storage.

4. Conclusions

Salt water reduces the time that the mixture is in contact with frozen water, which, as corroborated by the results, is one of the most harmful effects for the mixture.

When the bituminous mixture is subjected to freeze–thaw cycles, salt water has a protective effect on the specimens that remain submerged in it. The salt in the water protects the aggregate-binder adhesive, maintains the mechanical strength and increases the number of load cycles for any strain range.

The results of the Wheel Track Slope and the Rut Deep for reference mixtures and those that remained outdoors for one year are suitable for most climates zones and traffic.

The results of the Fatigue Test and Dynamic Modulus Test have no significant variation between the different series analysed.

The results show that, although the temperature has an injurious effect on the mechanical properties, the specimens submerged in salt water obtain better results than their analogs that are submerged in distilled water.

Author Contributions: Conceptualization, L.J.-G., Á.V.-Z. and M.Á.C.-P.; methodology, L.J.-G.; software, L.J.-G. and E.T.-L.-Z.; validation, L.J.-G., Á.V.-Z. and M.Á.C.-P.; formal analysis, L.J.-G.; investigation, L.J.-G; resources, L.J.-G.; data curation, L.J.-G. and E.T.-L.-Z.; writing—original draft preparation, L.J.-G.; writing—review and editing, L.J.-G., Á.V.-Z. and M.Á.C.-P.; supervision, L.J.-G., Á.V.-Z. and M.Á.C.-P. All authors have read and agreed to the published version of the manuscript.

Funding: This research received no external funding

Conflicts of Interest: The authors declare no conflict of interest.

References

1. Bäckström, M.; Karlsson, S.; Bäckman, L.; Folkeson, L.; Lind, B. Mobilisation of heavy metals by deicing salts in a roadside environment. *Water Res.* **2003**, *38*, 720–732. [CrossRef] [PubMed]
2. Norrström, A.C. Metal mobility by de-icing salt from an infiltration trench for highway runoff. *Appl. Geochem.* **2005**, *20*, 1907–1919. [CrossRef]
3. Engelsen, C.J.; Wibetoe, G.; van der Sloot, H.A.; Lund, W.; Petkovic, G. Field site leaching from recycled concrete aggregates applied as sub-base material in road construction. *Sci. Total Environ.* **2012**, *427–428*, 86–97. [CrossRef] [PubMed]
4. Norrström, A.C.; Jacks, G. Concentration and fractionation of heavy metals in roadside soils receiving de-icing salts. *Sci. Total Environ.* **1998**, *218*, 161–174. [CrossRef]
5. Rivett, M.O.; Cuthbert, M.O.; Gamble, R.; Connon, L.E.; Pearson, A.; Shepley, M.G.; Davis, J. Highway deicing salt dynamic runoff to surface water and subsequent infiltration to groundwater during severe UK winters. *Sci. Total Environ.* **2016**, *565*, 324–338. [CrossRef] [PubMed]
6. Snodgrass, J.W.; Moore, J.; Lev, S.M.; Casey, R.E.; Ownby, D.R.; Flora, R.F.; Izzo, G. Influence of Modern Stormwater Management Practices on Transport of Road Salt to Surface Waters. *Environm. Sci. Technol.* **2017**, *51*, 4165–4172. [CrossRef] [PubMed]
7. García, L. Eficiencia en vialidad invernal. Materiales empleados en los trabajos para el mantenimiento de la vialidad invernal. In Proceedings of the XII Jornadas de Conservación de Carreteras, Madrid, Spain, 10–11 November 2010.
8. Muthumani, A.; Fay, L.; Akin, M.; Wang, S.; Gong, J.; Shi, X. Correlating lab and field tests for evaluation of deicing and anti-icing chemicals: A review of potential approaches. *Cold Reg. Sci. Technol.* **2013**, *97*, 21–32. [CrossRef]
9. Shi, X.; Veneziano, D.; Xie, N.; Gong, J. Use of chloride-based ice control products for sustainable winter maintenance: A balanced perspective. *Cold Reg. Sci. Technol.* **2013**, *86*, 104–112. [CrossRef]
10. Klein-Paste, A.; Wåhlin, J. Wet pavement anti-icing—A physical mechanism. *Cold Reg. Sci. Technol.* **2013**, *96*, 1–7. [CrossRef]
11. Ikiz, N.; Galip, E. Computerized decision tree for anti-icing/pretreatment applications as a result of laboratory and field testings. *Cold Reg. Sci. Technol.* **2016**, *126*, 90–108. [CrossRef]
12. Trenouth, W.R.; Gharabaghi, B.; Perera, N. Road salt application planning tool for winter de-icing operations. *J. Hydrol.* **2015**, *524*, 401–410. [CrossRef]
13. Wang, D.; Xie, X.; Oeser, M.; Steinauer, B. Influence of the gritting material applied during the winter services on the asphalt surface performance. *Cold Reg. Sci. Technol.* **2015**, *112*, 39–44. [CrossRef]
14. Hassan, Y.; El Halim, A.O.A.; Razaqpur, A.G.; Bekheet, W.; Farha, M.H. Effects of Runway Deicers on Pavement Materials and Mixes: Comparison with Road Salt. *J. Transp. Eng.* **2012**, *128*, 385–391. [CrossRef]
15. Feng, D.; Yi, J.; Wang, D.; Chen, L. Impact of salt and freeze–thaw cycles on performance of asphalt mixtures in coastal frozen region of China. *Cold Reg. Sci. Technol.* **2010**, *62*, 31–41. [CrossRef]
16. Juli-Gándara, L.; Vega-Zamanillo, Á.; Calzada-Pérez, M.Á. Sodium chloride effect in the mechanical properties of the bituminous mixtures. *Cold Reg. Sci. Technol.* **2019**, *164*, 102776. [CrossRef]

17. Goh, S.W.; You, Z. Evaluation of Hot-Mix Asphalt Distress under Rapid Freeze-Thaw Cycles using Image Processing Technique. In Proceedings of the 12th International Conference of Transportation Professionals (CICTP 2012), Beijing, China, 3–6 August 2012; pp. 3305–3315.
18. Tarefder, R.; Faisal, H.; Barlas, G. Freeze-thaw effects on fatigue LIFE of hot mix asphalt and creep stiffness of asphalt binder. *Cold Reg. Sci. Technol.* **2018**, *153*, 197–207. [CrossRef]
19. Özgan, E.; Serin, S. Investigation of certain engineering characteristics of asphalt concrete exposed to freeze-thaw cycles. *Cold Reg. Sci. Technol.* **2013**, *85*, 131–136. [CrossRef]
20. Islam, M.R.; Tarefder, R.A. Effects of Large Freeze-Thaw Cycles on Stiffness and Tensile Strength of Asphalt Concrete. *J. Cold Reg. Eng.* **2016**, *30*, 06014006. [CrossRef]
21. Teltayev, B.B.; Rossi, C.O.; Izmailova, G.G.; Amirbayev, E.D. Effect of Freeze Thaw Cycles on Mechanical Characteristics of Bitumen and Stone Mastic Asphalts. *Appl. Sci.* **2019**, *9*, 458. [CrossRef]
22. UNE-EN 1097-2:2010. *Tests for Mechanical and Physical Properties of aggregaTes—Part 2: Methods for the Determination of Resistance to Fragmentation*; UNE: Madrid, Spain, 2010.
23. UNE-EN 1097-6:2014. *Tests for Mechanical and Physical Properties of Aggregates—Part 6: Determination of Particle Density and Water Absorption*; UNE: Madrid, Spain, 2014.
24. UNE-EN 933-3:2012. *Tests for Geometrical Properties of Aggregates—Part 3: Determination of Particle Shape—Flakiness Index*; UNE: Madrid, Spain, 2012.
25. UNE-EN 1426:2015. *Bitumen and Bituminous Binders. Determination of Needle Penetration*; UNE: Madrid, Spain, 2015.
26. UNE-EN 1427:2015. *Bitumen and Bituminous Binders. Determination of the Softening Point. Ring and BALL Method*; UNE: Madrid, Spain, 2015.
27. UNE-EN 12593:2015. *Bitumen and Bituminous Binders. Determination of the Fraass Breaking Point*; UNE: Madrid, Spain, 2015.
28. UNE-EN 12697-34:2013. *Bituminous Mixtures—Test Methods for Hot Asphalt—Part 34: Marshall Test*; UNE: Madrid, Spain, 2013.
29. Pliego de Prescripciones Técnicas Generales para Obras de Carreteras y Puentes (PG-3). 2017. Available online: http://www.arquitectosdecadiz.com/wp-content/uploads/2017/12/PG3.pdf (accessed on 20 January 2020).
30. UNE-EN 12697-12:2009. *Bituminous Mixtures—Test Methods for Hot Mix Asphalt—Part 12: Determination of the Water Sensitivity of Bituminous Specimens*; UNE: Madrid, Spain, 2009.
31. UNE-EN 12697-30:2013. *Bituminous Mixtures—Test Methods for Hot Mix Asphalt—Part 30: Specimen Preparation by Impact Compactor*; UNE: Madrid, Spain, 2013.
32. UNE-EN 12697-22:2008+A1:2008. *Bituminous Mixtures—Test Methods for Hot Mix Asphalt—Part 22: Wheel Tracking*; UNE: Madrid, Spain, 2008.
33. UNE-EN 12697-33:2006+A1:2007. *Bituminous Mixtures—Test Methods for Hot Mix Asphalt—Part 33: Specimen Prepared by Roller Compactor*; UNE: Madrid, Spain, 2007.
34. UNE-EN 12697-26:2012. *Bituminous Mixtures—Test Methods for Hot Mix Asphalt—Part 26: Stiffness*; UNE: Madrid, Spain, 2012.
35. UNE-EN 12697-24:2013. *Bituminous Mixtures—Test Methods for Hot Mix Asphalt—Part 24: Resistance to Fatigue*; UNE: Madrid, Spain, 2013.
36. Aleksandrov, A.A.; Dzhuraeva, E.V.; Utenkov, V.F. Thermal Conductivity of Sodium Chloride Aqueous Solutions. *Therm. Eng.* **2013**, *60*, 190–194. [CrossRef]

© 2020 by the authors. Licensee MDPI, Basel, Switzerland. This article is an open access article distributed under the terms and conditions of the Creative Commons Attribution (CC BY) license (http://creativecommons.org/licenses/by/4.0/).

Article

MSWI Bottom Ash Application to Resist Sulfate Attack on Concrete

Yongzhen Cheng *, Yun Dong, Jiakang Diao, Guoying Zhang, Chao Chen and Danxi Wu

Faculty of Architecture and Civil Engineering, Huaiyin Institute of Technology, Huai'an 223001, China; dyunhyit@hyit.edu.cn (Y.D.); 201861228007@njtech.edu.cn (J.D.); 18351888629@163.com (G.Z.); 201861228005@njtech.edu.cn (C.C.); 201861228010@njtech.edu.cn (D.W.)
* Correspondence: 230139226@seu.edu.cn

Received: 28 October 2019; Accepted: 21 November 2019; Published: 25 November 2019

Abstract: This research provides a strategy for partially replacing cement with municipal solid waste incineration (MSWI) bottom ash (BA) to improve the performance of concrete against sulphate attack. Mortar strength tests were performed firstly to evaluate the hydration activity of the ground BA. Concrete specimens were cured in standard conditions and immersed in a solution that contained 10% sodium sulfate. Then, the compressive strength of these specimens was measured to investigate the mechanical properties and durability of the concrete. Next, the capillary porosity of the concrete was determined from the volume fractions of water lost in specimens. Finally, the transport of the sulphate solution in concrete was analyzed using capillary rise, crystallization rate, and solution absorption tests. The results indicated that BA had a certain hydration activity. The equivalent replacement of cement by BA decreased the compressive strength of the specimens but increased the durability of the concrete. There was an excellent correlation between capillary rise height, sulfate solution absorption amount, crystallization rate, and coarse capillary porosity. The addition of BA can decrease the coarse capillary porosity and further slow the capillary transport and crystallization of sulfate solution in concrete. Overall, the replacement of cement with BA can improve the durability of concrete and actualize the utilization of MSWI residues as a resource.

Keywords: MSWI bottom ash; concrete; sulfate attack; capillary transport; crystallization

---

1. Introduction

The quantity of municipal solid waste continues to increase; it exceeded 2.1 billion tons as of 2017 in China, but it is still growing at a rate of 5% to 8% a year [1]. At present, incineration is the most effective way to realize the reduction and reutilization of municipal solid waste and render it harmless [2–4]. Incineration decreases waste quality by 70% and volume by up to 90% [5]. Thus, the purpose of waste reduction has been initially achieved. In addition, it generates energy from thermal combustion [6]. Waste incineration produces a large amount of bottom ash and fly ash. Bottom ash (BA) refers to the residue discharged from the end of the hearth, which is the main component of ash residue and is close to 80% of the total weight of bottom ash and fly ash. Therefore, further harmless treatment of municipal solid waste incineration (MSWI) bottom ash and its utilization as a resource are still urgent problems to be solved.

Cement concrete is the best-known type of construction material. Due to its good construction performance, economy, and durability, cement concrete has widely been used in structures such as buildings, bridges, and tunnels [7]. However, it has been found that concrete does not last as long as expected. Many destructive factors, such as sulphate attack, chloride ion penetration, carbonation, and more, can decrease the durability of concrete [8–10]. The reasons for damage to the durability of concrete due to sulfate attack are divided into two major categories, namely, physical attack and chemical corrosion [11]. Chemical corrosion is mainly caused by chemical reaction between the salt solution and the hydration products of the cement [12]. Physical attack refers to the destruction of concrete by the crystallization of the salt solution [13]. This destruction comes from volume expansion after the salt crystallization. The resistance of concrete to sulfate attack is mainly affected by the content of tricalcium aluminate, the amount of cement, and its compactness. The addition of industrial mineral admixtures can effectively improve the performance of concrete against sulphate attack [14–16]. After cement is partly replaced by coal ash, silica fume, and slag, and the amount of cement in the concrete is reduced accordingly, the content of calcium aluminate is relatively reduced. Furthermore, compared with cement, such industrial mineral admixtures have lower fineness and particle size; therefore, concrete admixed with these admixtures has greater compactness. Theoretically, concrete mixed with a mineral admixture has better resistance to sulfate attack.

In this study, a strategy was proposed to improve the sulfate resistance of concrete by mixing it with MSWI bottom ash. This proposal was also based on the chemical constitution of MSWI bottom ash, which is typically rich in silica and calcium oxide with minimum amounts of heavy metals, as it is classified as nonhazardous waste by the China Hazardous Waste List. Consequently, it is available to be reused as a secondary building material [17–19]. Many studies have strongly stated that MSWI bottom ash has some hydration activity and can be used to manufacture mortar [20,21]. However, the use of MSWI bottom ash to resist sulfate attack on concrete has been rarely reported, and the transportation and crystallization processes of salt solution in concrete remain unknown. Herein, detailed research was conducted by means of strength analysis, porosity measurements, and capillary rise and crystallization tests. Various water/cement ratios (W/C) and feedings (cement/BA) were found to have great impacts on the mechanical properties and salt solution transportation.

2. Materials and Methods

2.1. Materials

2.1.1. MSWI Bottom Ash

The MSWI bottom ash used in this research was sampled from a waste incineration power plant in Huai'an, China. The impurities and heavy metal ions in the samples were removed by washing treatment and magnetic separation. Ball-milling was conducted on an XQM-8 variable-frequency planetary ball mill to reduce the particle size to a maximum of 180 μm. Figure 1 shows the original and ground MSWI bottom ash. The particle size distribution of the BA was determined using a Bettersize 2000 laser particle analyzer, and the result is presented in Figure 2. Meanwhile a larger specific surface area of 6631 cm^2/g was obtained.

Figure 1. Municipal solid waste incineration (MSWI) bottom ash before (**a**) and after (**b**) ball milling.

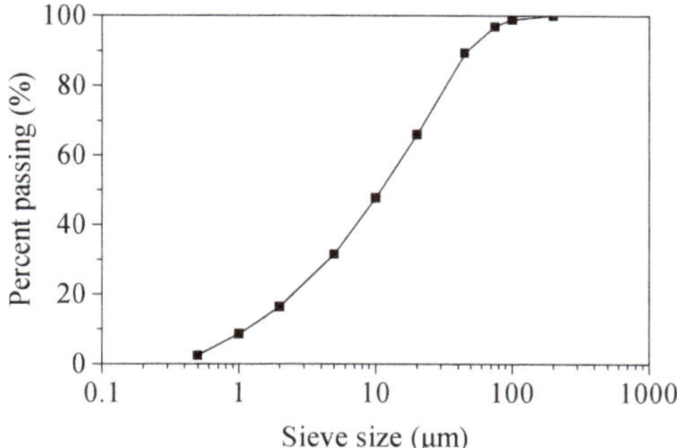

Figure 2. Particle size distribution for bottom ash (BA).

X-ray fluorescence spectroscopy (XRF) was performed using a Philips PW2400 instrument to determine the chemical composition of the MSWI bottom ash. The major, minor, and trace elements in the BA are presented in Table 1. Si, Al, Fe, and Ca together accounted for 68.59% of the total elements. Thus, the BA had a similar chemical composition to Portland cement. It belongs to a typical chemical system of $CaO–SiO_2–Al_2O_3–Fe_2O_3$ and should have hydration activity.

Table 1. The chemical (oxide) composition of BA.

Oxides	wt %
SiO_2	48.41
CaO	14.78
Al_2O_3	11.99
Na_2O	3.25
Fe_2O_3	5.40
SO_3	1.86
K_2O	1.42
MgO	1.78
TiO_2	0.76

2.1.2. Cement

Ordinary Portland cement with a grade of 42.5 was used in this study. The physical and mechanical properties of the cement measured in accordance with the Chinese standards are presented in Table 2.

Table 2. Material properties of cement.

Property	Requirements		Test Result	Test Method
Soundness	Qualified	GB 175-2007 [22]	Qualified	T0505-2005 (JTG E30-2005) [23]
Initial setting time (min)	$\nless 45$		97	T0505-2005 (JTG E30-2005) [23]
Final setting time (min)	$\ngtr 390$		180	
Compressive strength (MPa)	3d $\nless 17$		28.6	T0506-2005 (JTG E30-2005) [23]
	28d $\nless 42.5$		57.6	
Flexural strength (MPa)	3d $\nless 3.5$		4.5	
	28d $\nless 6.5$		9.0	
Specific area (cm^2/g)	3000		3550	T 050-2005 (JTG E30-2005) [23]

2.1.3. Natural Sand and Crushed Stone

The natural sand and crushed stone used in this research were sampled from a construction site in Huai'an, China. Sieve testing was performed in accordance with JCJ 52-2006 [24]. The grain size distribution curves of the natural sand and crushed stone are presented in Figures 3 and 4, respectively. The flat–elongated particle content, crushing value, clay content, and other parameters were determined in accordance with Chinese standards, and the test results are presented in Table 3. The water used for manufacturing the mortar and concrete in this research was local tap water.

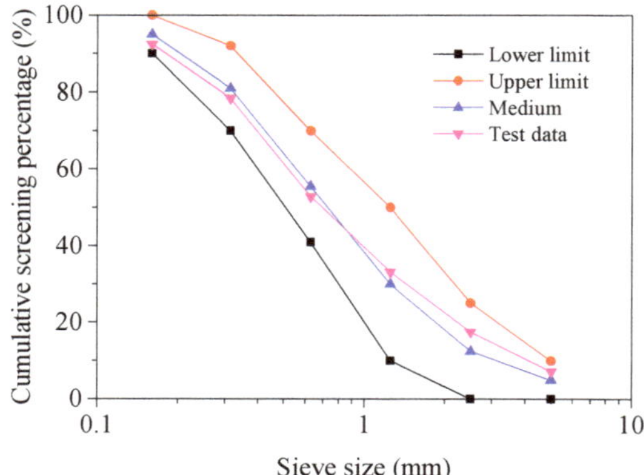

Figure 3. Grain size distribution curve of natural sand.

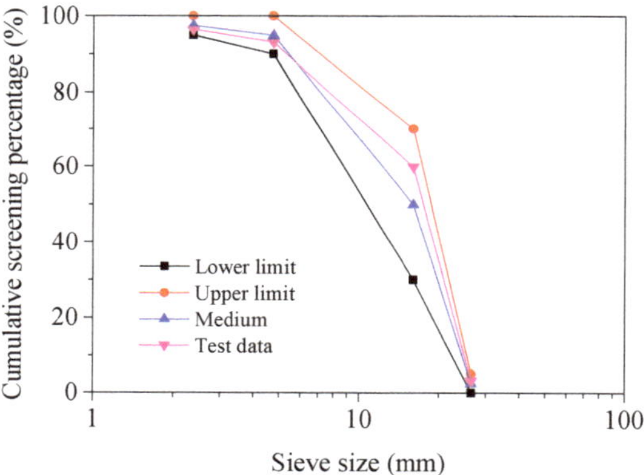

Figure 4. Grain size distribution curve of crushed stone.

Table 3. Material properties of natural sand and crushed stone.

Property	Natural Sand		Crushed Stone		Standard
	Requirements	**Test Result**	**Requirements**	**Test Result**	
Fineness modulus	NA	2.57	—	—	JCJ 52-2006 [24]
Flat–elongated particles (wt %)	—	—	$\not> 15$	13.5	
Clay content (wt %)	$\not> 3.0$	2.5	$\not> 1.0$	0.8	
Clay lump content (wt %)	$\not> 1.0$	0.8	$\not> 0.5$	0.3	
Crushing value (wt %)	—	—	$\not> 16$	12.8	
Ruggedness (wt %)	$\not> 8$	6	$\not> 8$	5	
Apparent density (g/cm^3)	NA	2.586	NA	2.714	

2.1.4. Mix Proportion Design

The mix proportion of mortar, determined according to GB/T 17671-199 [25], is shown in Table 4. The mix ratio of materials in mortar should be one cement, three standard sand, and one-half water by weight. Here, standard sand refers to natural quartz sea sand with SiO_2 content greater than 96%. After washing and sieving, it was processed into standard sand that meets the ISO requirements.

Concrete with a strength grade of C30 was prepared in the laboratory. The mixture ratio design of the concrete was carried out in accordance with JGJ 55-2000 [26]. Firstly, the mixed strength of the concrete was determined according to the strength grade of the designed concrete. Then, the water–cement ratio (W/C) was calculated based on the mixed strength of the concrete and strength value of the cement. Next, according to the slump and particle size of the stone, the water consumption was determined, and the amount of cementing material was calculated. Finally, the sand ratio was selected, and the amount of sand and crushed stone was calculated according to the maximum particle size of the stone and fineness modulus of the sand.

The polycarboxylic water-reducing agent mixed into the concrete accounted for 0.8% of the cementing material by volume. The water reducing rate was up to 29%. The final concrete mix proportions are shown in Table 4.

Table 4. Mix proportions of materials in the mortar and concrete.

Material	Cement Mortar	Cement Concrete
Cement (kg/m^3)	450	330
Water (kg/m^3)	225	130
Natural sand (kg/m^3)	1350	614
Crushed stone (kg/m^3)	—	1248
Water-cement ratio (W/C)	0.5	0.4

2.2. Combination Scheme

Laboratory tests were performed to evaluate the effect of BA content on concrete under sulphate attack. In addition, strength tests were conducted on the mortar specimens to evaluate the hydration activity of BA. The combination scheme of cementitious materials for mortar and concrete is presented in Table 5. In this study, five groups of mortar and concrete samples with BA content from 10% to 30% were prepared. In addition, a group of samples without BA was prepared. Each group consisted of three specimens.

Table 5. Cementitious material combination scheme for mortar and concrete.

Material	Cement (kg/m^3)	BA (kg/m^3)	Material	Cement (kg/m^3)	BA (kg/m^3)	Cement/BA	Designation
Cement mortar	405	45	Cement concrete	297	33	90:10	C90BA10
	382.5	67.5		280.5	49.5	85:15	C85BA15
	360	90		264	66	80:20	C80BA20
	337.5	112.5		247.5	82.5	75:25	C75BA25
	315	135		231	99	70:30	C70BA30

2.3. Experimental Methods

2.3.1. Mechanical Property Measurements

The specimen preparation and strength tests of the mortar were conducted in accordance with GB/T 17671-199 [25]. The well-stirred mixtures of cement, BA, sand, and water were put into a mold (40 × 40 × 160 mm) fixed on a vibrating table; after vibrating 120 times, the mixtures together with the mold were stored in a curing room (maintained at 20 ± 2 °C and no less than 95% RH). Then, form stripping was carried out. After 28 days of curing in standard conditions, the flexural and compressive strengths of the mortar were determined using a bending and compression tester.

The activity index of BA was determined with reference to GB T1596-2017 [27]. According to this standard, the compressive strengths of mortar with 30% and without fly ash were tested under a water-to-binder ratio of 0.5. Here, the activity index is defined as the compressive strength of the mortar with BA divided by the compressive strength of the mortar without BA, and it can be calculated as follows:

$$H = \frac{R}{R_0} \times 100 \tag{1}$$

where H is the activity index (%); R is the compressive strength of mortar with BA at 28 days of curing (MPa); and R_0 is the compressive strength of mortar without BA at 28 days of curing (MPa).

The concrete specimens were prepared and cured according to GB/T 50081-2002 [28]. All the materials were accurately weighed and put into the mixing pan. After stirring, specimens $100 \times 100 \times 100$ mm in size were manufactured by the vibration molding method. Next, all specimens were left to stand for 24 h at a temperature of $20 \pm 5\ °C$. In order to evaluate the strength property of the concrete, the specimens were cured in a curing room (maintained at $20 \pm 2\ °C$ and no less than 95% RH). After 28 days of curing, the compressive strength of the concrete was determined using a WAW-B Electro-hydraulic universal testing machine.

For durability tests, specimens $100 \times 100 \times 100$ mm in size were divided into two series. The first one was cured under standard curing conditions. The second one was firstly cured at T = $20 \pm 2\ °C$ and RH $\geq$ 95% for 28 days and then totally immersed in a solution containing 10% sodium sulfate for 60 days; the solutions were renewed monthly. The sulfate damage was evaluated mechanically by determining the compressive strength loss of the specimens using the following equation:

$$Strength\ loss\ (\%) = \frac{R_1 - R_2}{R_1} \times 100 \tag{2}$$

where R_1 is the compressive strength of concrete specimens under standard curing conditions (MPa) and R_2 is the compressive strength of concrete specimens in sulfate solutions (MPa).

2.3.2. Concrete Porosity Measurements

The concrete porosity was determined in accordance with the standard test method [29]. The porosity of concrete can be obtained indirectly from the water loss rate of saturated concrete specimens under certain conditions. The concrete specimens were prepared and cured for 28 days in standard conditions. After vacuum saturation, the water on the specimen surface was wiped off using a dry cloth. The mass of the specimen, measured using an electronic balance, was noted down. Then the specimen was cured at RH = 90% for 30 days. When water diffusion in the concrete reached equilibrium states, the mass of the same specimen was measured again. Next, the specimen was oven-dried to a constant weight at T = 105 °C. The specimen was finally weighed after cooling. The concrete porosity was calculated using the following equations:

$$P_{fine} = \frac{(M_0 - M_1) \times \rho_C}{M_0 \times \rho_w} \times 100\% \tag{3}$$

$$P_{total} = \frac{(M_0 - M_2) \times \rho_C}{M_0 \times \rho_w} \times 100\% \tag{4}$$

$$P_{coarse} = P_{total} - P_{fine} \tag{5}$$

where P_{total} is the total porosity of the concrete specimen (%); P_{fine} is the fine capillary porosity of the concrete specimen (%); P_{coarse} is the coarse capillary porosity of the concrete specimen (%); M_0 is the mass of the saturated concrete specimen (kg); M_1 is the mass of the concrete specimen cured at RH = 90% for 30 days (kg); M_2 is the mass of the dried concrete specimen; ρ_c is the density of the concrete (kg/m^3); and ρ_w is the density of the water (kg/m^3).

2.3.3. Capillary Rise and Crystallization Tests

A concrete specimen $150 \times 150 \times 150$ mm in size was prepared and cured by following the standard method. After 28 days of curing, a cylinder (100 mm in diameter and 150 mm in length) was drilled from the specimen (Figure 5a). The cylinder part was oven-dried to a constant weight at T = 60 °C. Then, this part was partially immersed in water to measure the rising height of capillary water in the concrete; the test method is shown in Figure 5c. The hollow part was used to measure the capillary crystallization rate of sodium sulfate solution in the concrete. A schematic diagram of the capillary crystallization tests is shown in Figure 5b. The bottom surface was firstly sealed. Then, a solution containing 5% sodium sulfate was poured into the cavity. The crystallization of sodium sulfate solution from the side wall and the thickness of concrete wall were observed and recorded every half hour. The above tests were performed at T = 20 ± 2 °C and RH = 60% ± 5%. The capillary crystallization rate was calculated using the following equation:

$$V_s = L/T \tag{6}$$

where V_s is the capillary crystallization rate (cm/s); L is the thickness of the sidewall (cm); and T is the time of original crystallization on the concrete surface (s).

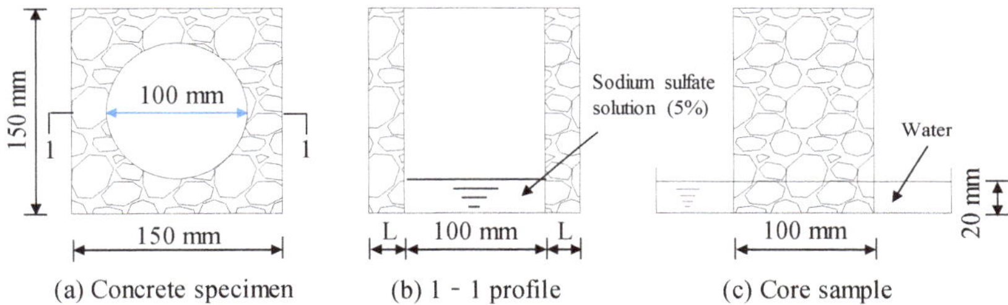

Figure 5. Schematic diagram of the capillary rise and crystallization tests (**a**) Concrete specimen without core sample; (**b**) 1-1 Profile of Figure 5a; (**c**) Core sample.

2.3.4. Solution Absorption Measurements

A specimen (100 mm in diameter and 100 mm in length) was prepared and cured for 28 days. Next, the solution absorption was measured in the laboratory; a schematic diagram is shown in Figure 6. All the side faces of the concrete specimen were sealed using epoxy resin. The specimen was oven-dried to a constant weight at T = 60 °C for no less than 12 h and was weighed after cooling. Then, the specimen was placed in a solution containing 5% sodium sulfate, keeping the water surface 5 mm above the bottom surface. The solution temperature was kept constant at 20 °C. The water on the specimen surface was wiped off using a dry cloth, and the weight of the specimen was measured at regular intervals.

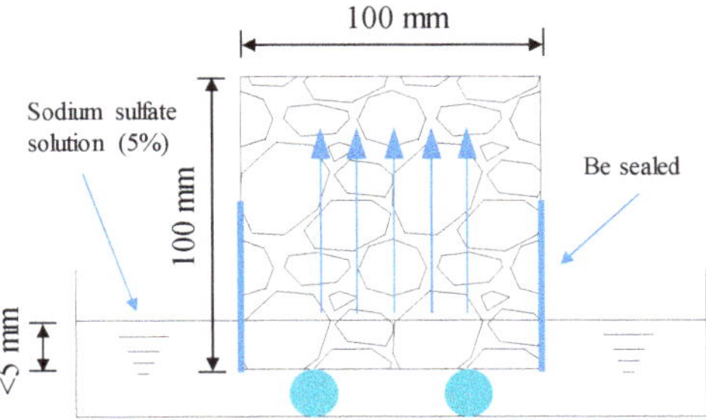

Figure 6. Schematic diagram of solution absorption measurements.

3. Results and Discussion

3.1. Strength Properties of Mortar and Concrete

The flexural and compressive strengths of the mortar samples with various contents of BA at 28 days of curing are shown in Table 6. Both the compressive and flexural strengths of the mortar decreased with increasing BA content, which is similar to the results of previous research [30–32]. The compressive strength of mortar showed a large decrease at the addition of BA from 10% to 20%. There was not a significant decrease in compressive strength at the addition of BA from 20% to 25%. Beyond 25%, the compressive strength decreased significantly again. Similar to the compressive strength, with increasing BA, the flexural strength also showed a rapid reduction first and then a gentle reduction, followed by another significant reduction.

Low hydration activity of BA was obtained, and the activity index was only 43% with the addition of 30% BA. The hydration activity of BA was apparently smaller in comparison with the records in the literature [33]. In the literature, a low W/C value of 0.38 was used. However, a similar industrial mineral admixture, electric arc furnace dust, has high hydration activity with W/C values from 0.35 to 0.7 [34]. Hence, the W/C of 0.5 used in this study is not the main cause of the low hydration activity. The samples in previous research were prepared by melting the MSWI fly ash at a high temperature and then water-quenching [33]. In this study, the BA was prepared by artificially removing the impurities. In addition, the particle size of the BA sample was controlled under 180 μm, making it hard to densify the microscopic structure of the mortar. In addition, our BA sample had a higher content of SiO_2 and a lower amount of CaO in comparison with the finer samples [35], and CaO may participate in the cement hydration process.

Table 6. Flexural strength and compressive strength of mortar samples with various contents of BA.

Samples	Flexural Strength (MPa)		Compressive Strength (MPa)		Activity Index (%)
	Average Value	Standard Deviation	Average Value	Standard Deviation	
Without BA	9.0	0.8	57.1	2.9	—
C90BA10	5.2	0.3	34.6	2.6	61
C85BA15	5.0	0.2	32.0	2.7	56
C80BA20	5.0	0.5	27.6	2.2	48
C75BA25	4.8	0.4	26.8	2.5	47
C70BA30	4.5	0.3	24.4	2.1	43

Figure 7 shows the compressive strength values of the concrete samples with different contents of BA at 28 days of curing. The compressive strength of the concrete decreased gradually with increasing addition of BA. In addition, the water/cement ratio (W/C) had a significant influence on the compressive strength: the larger the W/C, the smaller the compressive strength. The concrete samples with the addition of 10%, 15%, and 20% BA at a W/C of 0.35 met the strength requirement of C30, at 37 MPa, 34 MPa, and 32 MPa, respectively. When the W/C was 0.40, only the concrete samples with the addition of 10% and 15% BA met the strength requirement, at 35 MPa and 31 MPa, respectively. Unfortunately, the compressive strengths of all concrete specimens at a W/C of 0.35 were less than 30 MPa. This shows that the addition of BA could not improve the strength performance of the concrete due to its relatively low hydration activity.

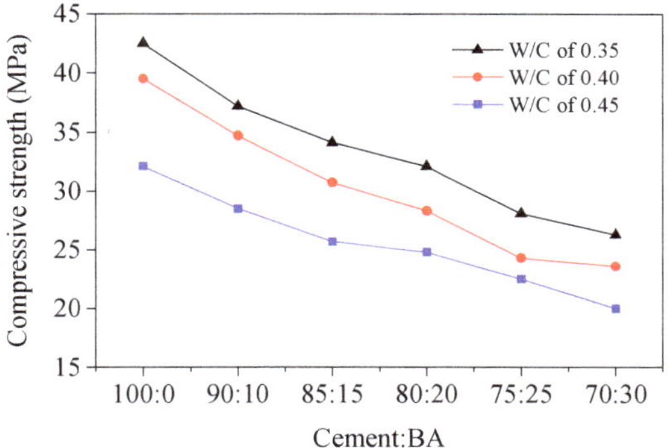

Figure 7. Influence of BA content on the compressive strength of concrete.

3.2. Concrete Durability

Figure 8 shows the compressive strength values of concrete specimens cured in standard conditions and immersed in sulfate solutions, also giving the strength loss. All the concrete specimens were manufactured with a W/C of 0.4. The compressive strengths of the concrete samples with the addition of 10%, 20%, and 30% BA decreased from 30.5 MPa, 27.7 MPa, and 26.0 MPa to 25.7 MPa, 25.6 MPa, and 24.9 MPa, respectively. Meanwhile the strength loss decreased from 15.6% to 7.5%, then to 4.2%. Thus, the more BA is added to the concrete, the smaller the strength loss will be. The BA can thus improve the performance of concrete against sulphate attack. To some extent, this is due to the very high surface area of the analyzed BA, which can fill in the pores of the concrete and prevent the sulphate solution from seeping into concrete. A similar phenomenon was found in cement concrete mixed with fly ash and other recycled micro powders [36–38].

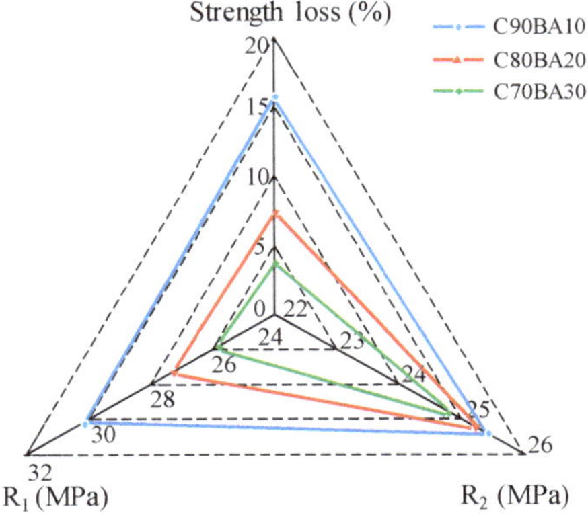

Figure 8. Strength loss in Na$_2$SO$_4$ solution.

The porosity values of cement concrete samples at different W/C and addition levels of BA are shown in Figure 9. The total porosity, coarse capillary porosity, and fine capillary porosity all increased with increasing W/C. The addition of BA was able to efficiently decrease the porosity of the concrete, especially the coarse capillary porosity. The pore structure and porosity are the key factors that affect the strength of cement-based materials [39]. They are also the decisive factors in the resistance of cement-based materials to invasive media [40]. The attack resistance of cement-based materials is poor when the porosity is large and the pores are interconnected. On the contrary, better performance of concrete against attack can be obtained. This explains why the concrete with BA had better sulfate resistance.

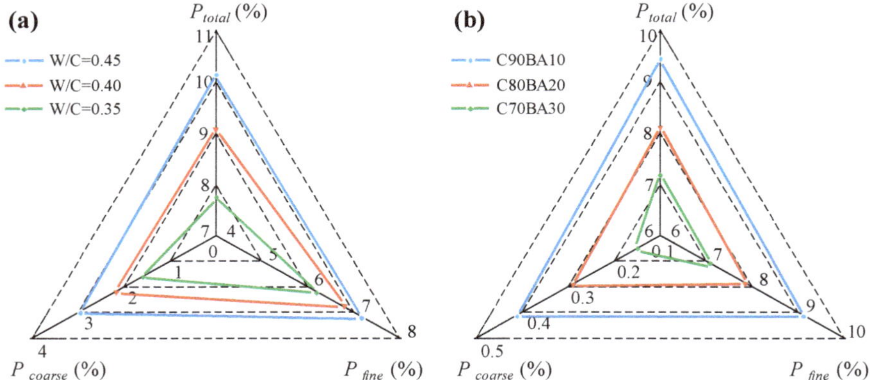

Figure 9. Influence of (**a**) W/C and (**b**) BA content on porosity of cement concrete.

3.3. Intrusion of Concrete by Sodium Sulfate Solution

Figure 10 shows the heights of capillary rise with time at different W/C values and BA contents. The height of capillary rise increased with increasing W/C. The capillary rise height of the concrete with BA was obviously lower than that of the concrete without BA.

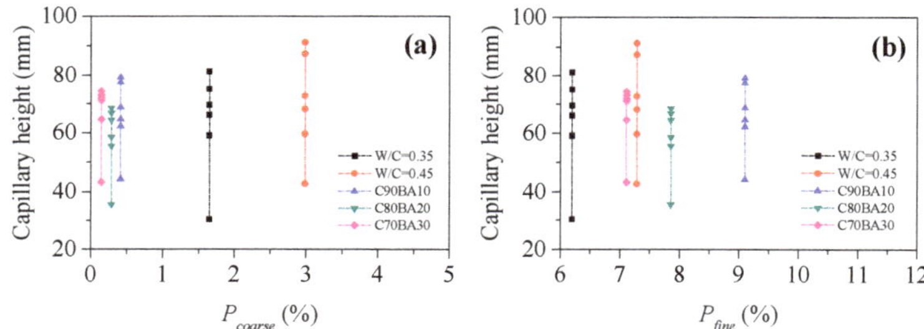

Figure 10. Height of capillary rise at (**a**) coarse and (**b**) fine capillary porosity.

The correlations of the capillary rise height versus the coarse and fine capillary porosity are shown in Figure 11. There was a good correlation between the height of capillary rise and the coarse capillary porosity (pore size of ≥30 nm) of the concrete, with a correlation coefficient of 0.819. However, the height of the capillary rise had little correlation with the fine capillary porosity (pore size of <30 nm), with a correlation coefficient of only 0.0669. Thus, the capillary porosity types with various pore sizes have different effects on capillary rise in concrete. The coarse capillary porosity plays a key role in the capillary transport of a solution in concrete [41].

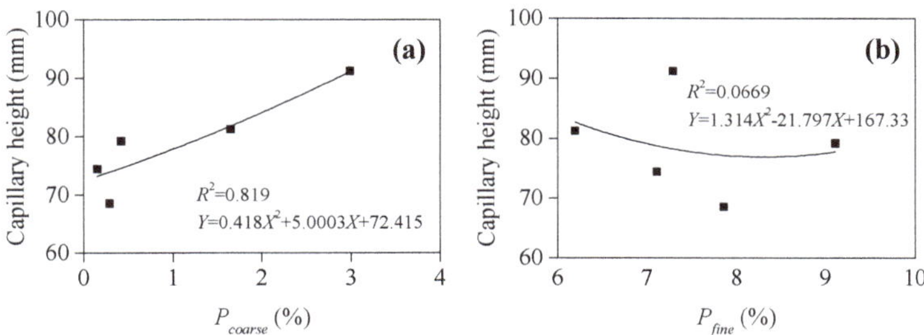

Figure 11. Correlations of capillary rise height vs. (**a**) coarse and (**b**) fine porosity.

Figure 12 shows the correlations of the crystallization rate of capillary transmission versus the coarse and fine porosity. The crystallization rate of capillary transmission increased with increasing W/C. A shorter time for sodium sulfate solution to reach the surface of concrete results in a larger area of crystallization on the concrete surface. The crystallization rate of the concrete with BA decreased significantly. In addition, the crystallization rate decreased with increasing BA content. There was a good correlation between the crystallization rate of capillary transmission and the coarse capillary porosity, with a correlation coefficient of 0.959. However, the crystallization rate had little correlation with the fine capillary porosity, with a correlation coefficient of only 0.0973. That also fully explained the importance of the coarse capillary porosity in the capillary transport of a solution in concrete.

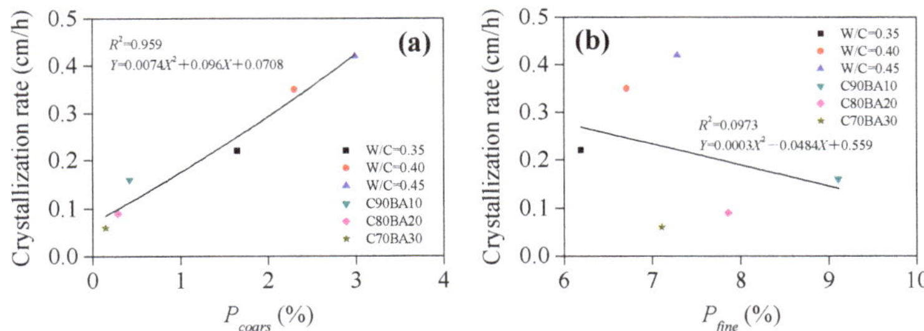

Figure 12. Correlations of crystallization rate vs. (**a**) coarse and (**b**) fine porosity.

The influence of the coarse and fine porosity on the mass of sodium sulfate solution absorption and their correlations are shown in Figure 13. W/C had a significant influence on the mass of sodium sulfate solution absorption, which was largest at a W/C of 0.45, followed by a W/C of 0.35. After the addition of BA, the absorption amount of sodium sulfate solution in the concrete decreased significantly, and it deceased with increasing BA content. There was a good correlation between the absorption amount of sodium sulfate solution and the coarse capillary porosity, with a correlation coefficient of 0.9793. However, a poor correlation between the absorption amount of sodium sulfate solution and the fine capillary porosity was found.

In conclusion, the fine capillary porosity has little influence on the height of capillary rise, sodium sulfate solution absorption, and crystallization rate of capillary transmission, while the coarse capillary porosity plays a key role in the intrusion of sodium sulfate solution into concrete. The concrete samples at various W/C values and BA contents had different porosity properties. Thus, W/C and BA content have an obvious influence on the transfer and crystallization of salt solution in concrete. The height of capillary rise, sodium sulfate solution absorption, and crystallization rate decreased with increasing BA content. All this proves that the addition of BA can improve the performance of concrete against sulphate attack.

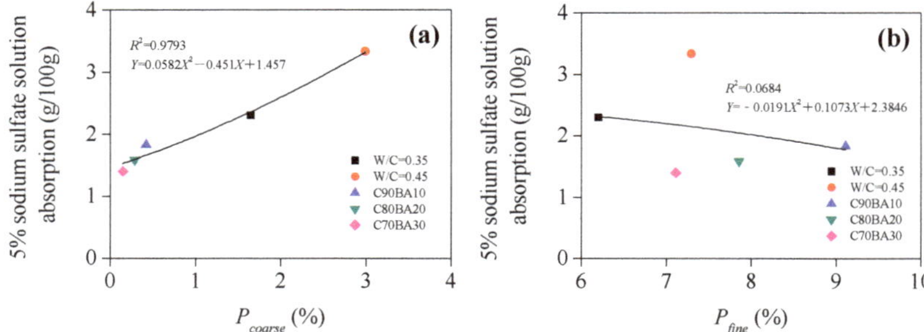

Figure 13. Correlations of mass of sodium sulfate solution absorption vs. (**a**) coarse and (**b**) fine porosity.

4. Conclusions

In this research, the durability of concrete with added BA was evaluated by immersing specimens in a solution containing 10% sodium sulfate. Then, capillary rise, crystallization rate, and solution absorption tests were performed to investigate the intrusion of concrete by sodium sulfate solution. Moreover, the crucial type of porosity was found with the help of porosity measurements. The following conclusions can be drawn:

(1)   BA has some hydration activity due to its chemical constitution. Even if 30% of the cement is replaced with BA, the compressive and flexural strength values are still greater than 42% and 50%, respectively, of those of the mortar without BA.

(2)   The compressive strength of the concrete decreased with increasing BA content; this is due to the weak pozzolanic reactivity of BA. However, BA can improve the performance of concrete against sulphate attack because BA has a very high surface area and fills in the pores of concrete.

(3)   The coarse capillary porosity plays a key role in the capillary transport and crystallization of sulfate solution in concrete.

(4)   The W/C and BA content have a certain impact on the porosity of concrete and further affect the capillary height, absorption amount, and crystallization rate of sulfate solution in concrete.

Furthermore, due to BA adjusting the strength and durability of the concrete, the optimum range of BA in concrete is from 10% to 15%. In this range, BA can not only improve the durability of concrete but also ensure the strength of concrete to meet design requirements.

The deficiency of this study is that no suitable treatment measures were found to enhance the hydration activity of BA. The objective of future research studies in this area is to improve the activity of BA by wet and high-temperature treatment. On this basis, the effect of BA on the strength and durability of concrete can be studied.

Author Contributions: Y.C. proposed the research method for analyzing the intrusion of sodium sulfate solution to reveal the durability of concrete. The experiments were performed and the manuscript was written by J.D., G.Z., C.C. and D.W. Y.D. proposed revision suggestions for this manuscript.

Funding: This research was funded by the Construction System Science and Technology Project in Jiangsu, China (Grant No. 2018ZD045), and the Training Program on Entrepreneurship and Innovation for University Students (Grant No. 2019JGYJ003 and 201911049023Y).

Conflicts of Interest: The authors declare that there are no conflicts of interest regarding the publication of this paper.

References

1.   National Bureau of Statistics of China. *China Statistical Yearbook*; China Statistics Press: Beijing, China, 2018.

2.   Loginova, E.; Volkov, D.S.; van de Wouw, P.M.F.; Florea, M.V.A.; Brouwers, H.J.H. Detailed characterization of particle size fractions of municipal solid waste incineration bottom ash. *J. Clean. Prod.* **2019**, *207*, 866–874. [CrossRef]

3. Trindade, A.B.; Palacio, J.C.E.; González, A.M.; Orozco, D.J.R.; Lora, E.E.S.; Renó, M.L.G.; del Olmo, O.A. Advanced exergy analysis and environmental assesment of the steam cycle of an incineration system of municipal solid waste with energy recovery. *Energy Convers. Manag.* **2018**, *157*, 195–214. [CrossRef]

4. Panepinto, D.; Zanetti, M.C. Municipal solid waste incineration plant: A multi-step approach to the evaluation of an energy-recovery configuration. *Waste Manag.* **2018**, *73*, 332–341. [CrossRef]

5. Del Valle-Zermeño, R.; Formosa, J.; Chimenos, J.M.; Martínez, M.; Fernández, A.I. Aggregate material formulated with MSWI bottom ash and APC fly ash for use as secondary building material. *Waste Manag.* **2013**, *33*, 621–627. [CrossRef] [PubMed]

6. Lam, C.H.K.; Ip, A.W.M.; Barford, J.P.; McKay, G. Use of incineration MSW ash: A review. *Sustainability* **2010**, *2*, 1943–1968. [CrossRef]

7. Gagg, C.R. Cement and concrete as an engineering material: An historic appraisal and case study analysis. *Eng. Fail. Anal.* **2014**, *40*, 114–140. [CrossRef]

8. Najjar, M.F.; Nehdi, M.L.; Soliman, A.M.; Azabi, T.M. Damage mechanisms of two-stage concrete exposed to chemical and physical sulfate attack. *Constr. Build. Mater.* **2017**, *137*, 141–152. [CrossRef]

9. Divsholi, B.S.; Lim, T.Y.D.; Teng, S. Durability properties and microstructure of ground granulated blast furnace slag cement concrete. *Int. J. Concr. Struct. Mater.* **2014**, *8*, 157–164. [CrossRef]

10. Duan, P.; Shui, Z.; Chen, W.; Shen, C. Enhancing microstructure and durability of concrete from ground granulated blast furnace slag and metakaolin as cement replacement materials. *J. Mater. Res. Technol.* **2013**, *2*, 52–59. [CrossRef]

11. Ragoug, R.; Metalssi, O.O.; Barberon, F.; Torrenti, J.M.; Roussel, N.; Divet, L.; de Lacaillerie, J.B.D.E. Durability of cement pastes exposed to external sulfate attack and leaching: Physical and chemical aspects. *Cem. Concr. Res.* **2019**, *116*, 134–145. [CrossRef]

12. Li, C.; Wu, M.; Chen, Q.; Jiang, Z. Chemical and mineralogical alterations of concrete subjected to chemical attacks in complex underground tunnel environments during 20–36 years. *Cem. Concr. Compos.* **2018**, *86*, 139–159. [CrossRef]

13. Bassuoni, M.T.; Rahman, M.M. Response of concrete to accelerated physical salt attack exposure. *Cem. Concr. Res.* **2016**, *79*, 395–408. [CrossRef]

14. Khan, R.A.; Ganesh, A. The effect of coal bottom ash (CBA) on mechanical and durability characteristics of concrete. *J. Build. Mater. Struct.* **2016**, *3*, 31–42.

15. Wang, D.; Zhou, X.; Meng, Y.; Chen, Z. Durability of concrete containing fly ash and silica fume against combined freezing-thawing and sulfate attack. *Constr. Build. Mater.* **2017**, *147*, 398–406. [CrossRef]

16. Bakharev, T.; Sanjayan, J.G.; Cheng, Y.B. Resistance of alkali-activated slag concrete to acid attack. *Cem. Concr. Res.* **2003**, *33*, 1607–1611. [CrossRef]

17. Polettini, A.; Pomi, R.; Sirini, P.; Testa, F. Properties of Portland cement-stabilised MSWI fly ashes. *J. Hazard. Mater.* **2001**, *88*, 123–138. [CrossRef]

18. Fernández Bertos, M.; Li, X.; Simons, S.J.R.; Hills, C.D.; Carey, P.J. Investigation of accelerated carbonation for the stabilisation of MSW incinerator ashes and the sequestration of CO_2. *Green Chem.* **2004**, *6*, 428–436. [CrossRef]

19. Chimenos, J.M.; Fernández, A.I.; Cervantes, A.; Miralles, L.; Fernández, M.A.; Espiell, F. Optimizing the APC residue washing process to minimize the release of chloride and heavy metals. *Waste Manag.* **2005**, *25*, 689–693. [CrossRef] [PubMed]

20. Saikia, N.; Mertens, G.; Van Balen, K.; Elsen, J.; Van Gerven, T.; Vandecasteele, C. Pre-treatment of municipal solid waste incineration (MSWI) bottom ash for utilisation in cement mortar. *Constr. Build. Mater.* **2015**, *96*, 76–85. [CrossRef]

21. Caprai, V.; Schollbach, K.; Brouwers, H.J.H. Influence of hydrothermal treatment on the mechanical and environmental performances of mortars including MSWI bottom ash. *Waste Manag.* **2018**, *78*, 639–648. [CrossRef]

22. General Administration of Quality Supervision, Inspection and Quarantine of the People's Republic of China. *Standardization Administration. Common Portland Cement (GB 175-2007)*; Standards Press of China: Beijing, China, 2007. (In Chinese)

23. Ministry of Transport of the People's Republic of China. *Test Methods of Cement and Concrete for Highway Engineering (JTG E30-2005)*; China Communications Press: Beijing, China, 2005. (In Chinese)

24. Ministry of Housing and Urban-Rural Development of the People's Republic of China. *Standard for Technical Requirements and Test Method of Sand and Crushed Stone (or Gravel) for Ordinary Concrete (JGJ 52-2006)*; China Building Industry Press: Beijing, China, 2006. (In Chinese)
25. The State Bureau of Quality and Technical Supervision. *Method of Testing Cement-Determination of Strength (GB/T 17671-1999)*; Standards Press of China: Beijing, China, 1999. (In Chinese)
26. Ministry of Housing and Urban-Rural Development of the People's Republic of China. *Specification for Mix Proportion Design of Ordinary Concrete (JGJ 55-2000)*; China Building Industry Press: Beijing, China, 2000. (In Chinese)
27. General Administration of Quality Supervision, Inspection and Quarantine of the People's Republic of China. *Fly Ash Used for Cement and Concrete (GB T1596-2017)*; Standards Press of China: Beijing, China, 2017. (In Chinese)
28. Ministry of Housing and Urban-Rural Development of the People's Republic of China; General Administration of Quality Supervision; Inspection and Quarantine of the People's Republic of China. *Standard for Test Method of Mechanical Properties on Ordinary Concrete (GB/T 50081-2002)*; China Building Industry Press: Beijing, China, 2002. (In Chinese)
29. Ngala, V.T.; Page, C.L. Effects of carbonation on pore structure and diffusional properties of hydrated cement pastes. *Cem. Concr. Res.* **1997**, *27*, 995–1007. [CrossRef]
30. Lin, K.L.; Chang, W.C.; Lin, D.F. Pozzolanic characteristics of pulverized incinerator bottom ash slag. *Constr. Build. Mater.* **2008**, *22*, 324–329. [CrossRef]
31. Li, X.G.; Lv, Y.; Ma, B.G.; Chen, Q.B.; Yin, X.B.; Jian, S.W. Utilization of municipal solid waste incineration bottom ash in blended cement. *J. Clean. Prod.* **2012**, *32*, 96–100. [CrossRef]
32. Tang, P.; Florea, M.V.A.; Spiesz, P.; Brouwers, H.J.H. Characteristics and application potential of municipal solid waste incineration (MSWI) bottom ashes from two waste-to-energy plants. *Constr. Build. Mater.* **2015**, *83*, 77–94. [CrossRef]
33. Lin, K.L.; Wang, K.S.; Lin, C.Y.; Lin, C.H. The hydration properties of pastes containing municipal solid waste incinerator fly ash slag. *J. Hazard. Mater.* **2004**, *109*, 173–181. [CrossRef]
34. Da Silva Magalhães, M.; Faleschini, F.; Pellegrino, C.; Brunelli, K. Cementing efficiency of electric arc furnace dust in mortars. *Constr. Build. Mater.* **2017**, *157*, 141–150. [CrossRef]
35. Tang, P.; Florea, M.V.A.; Spiesz, P.; Brouwers, H.J.H. Application of thermally activated municipal solid waste incineration (MSWI) bottom ash fines as binder substitute. *Cem. Concr. Compos.* **2016**, *70*, 194–205. [CrossRef]
36. Ramezanianpour, A.A.; Malhotra, V.M. Effect of curing on the compressive strength, resistance to chloride-ion penetration and porosity of concretes incorporating slag, fly ash or silica fume. *Cem. Concr. Compos.* **1995**, *17*, 125–133. [CrossRef]
37. Kalla, P.; Misra, A.; Gupta, R.C.; Csetenyi, L.; Gahlot, V.; Arora, A. Mechanical and durability studies on concrete containing wollastonite–fly ash combination. *Constr. Build. Mater.* **2013**, *40*, 1142–1150. [CrossRef]
38. Chindaprasirt, P.; Rukzon, S. Strength, porosity and corrosion resistance of ternary blend Portland cement, rice husk ash and fly ash mortar. *Constr. Build. Mater.* **2008**, *22*, 1601–1606. [CrossRef]
39. Chindaprasirt, P.; Jaturapitakkul, C.; Sinsiri, T. Effect of fly ash fineness on compressive strength and pore size of blended cement paste. *Cem. Concr. Compos.* **2005**, *27*, 425–428. [CrossRef]
40. Zakaria, M.; Cabrera, J.G. Performance and durability of concrete made with demolition waste and artificial fly ash-clay aggregates. *Waste Manag.* **1996**, *16*, 151–158. [CrossRef]
41. Xie, Y.J.; Ma, K.L.; Long, G.C.; Shi, M.X. Influence of mineral admixture on chlorid ion permeability of concrete. *J. Chin. Ceram. Soc.* **2006**, *34*, 1345. (In Chinese)

© 2019 by the authors. Licensee MDPI, Basel, Switzerland. This article is an open access article distributed under the terms and conditions of the Creative Commons Attribution (CC BY) license (http://creativecommons.org/licenses/by/4.0/).

Article

Thermal and Mechanical Properties of Cement Mortar Composite Containing Recycled Expanded Glass Aggregate and Nano Titanium Dioxide

Ali Yousefi [1], Waiching Tang [1,*], Mehrnoush Khavarian [1], Cheng Fang [2] and Shanyong Wang [3]

[1] School of Architecture and Built Environment, University of Newcastle, Callaghan, NSW 2308, Australia; ali.yousefi@newcastle.edu.au (A.Y.); mehrnoush.khavarian@newcastle.edu.au (M.K.)

[2] Global Centre for Environmental Remediation (GCER), University of Newcastle, Callaghan, NSW 2308, Australia; cheng.fang@newcastle.edu.au

[3] School of Civil Engineering, University of Newcastle, Callaghan, NSW 2308, Australia; shanyong.wang@newcastle.edu.au

* Correspondence: patrick.tang@newcastle.edu.au; Tel.: +61-249-217-246

Received: 14 February 2020; Accepted: 20 March 2020; Published: 26 March 2020

Abstract: One of the growing concerns in the construction industry is energy consumption and energy efficiency in residential buildings. Moreover, management of non-degradable solid glass wastes is becoming a critical issue worldwide. Accordingly, incorporation of recycled expanded glass aggregates (EGA) as a substitution for natural fine aggregate in cement composites would be a sustainable solution in terms of energy consumption in the buildings and waste management. This experimental research aims to investigate the effects of EGA on fresh and hardened properties and thermal insulating performance of cement mortar. To enhance the mechanical properties and water resistance of the EGA-mortar, nano titanium dioxide (nTiO$_2$) was used as nanofillers. The results showed an increase in workability and water absorption of the EGA-mortar. In addition, a significant decrease in bulk density and compressive strength observed by incorporating EGA into the cement mortar. The EGA-mortar exhibited a low heat transfer rate and excellent thermal insulation property. Furthermore, inclusion of nTiO$_2$ increased compressive strength and water resistance of EGA-mortar, however, their heat transfer rate was increased. The results demonstrated that EGA-mortar can be integrated into the building envelop or non-load bearing elements such as wall partition as a thermal resistance to reduce the energy consumption in residential buildings.

Keywords: industrial waste; sustainable concrete; recycled expanded glass

---

1. Introduction

In the last few decades, demand for energy consumption in the residential building has risen and there is a high intention for reducing energy consumption in the buildings. The energy efficiency of the buildings has become increasingly critical with the rising costs of energy as well as increasing awareness on global warming effects [1,2]. Furthermore, waste management has become a critical issue. In fact, non-degradable wastes such as glass are unable to break down naturally which is developing environmental problems [3]. In Australia, about 1.1 Mt of glass waste was generated in the year of 2016–2017 from that 43% was stockpiled. In New South Wales, companies accepting the landfill levy to dump their glass waste in landfill or arranging to relocate the waste to other states where the landfill levy does not apply. Thus, use of waste glass in the large scale is a sustainable solution in terms of reduction of carbon footprint and saving the costs and energy. The construction industry is a potential sector for utilization of waste glass. In this regard, use of solid wastes for manufacturing the building materials with high thermal insulation properties is an effective approach toward sustainable development and decreasing the energy consumption in buildings [4–6]. It has been reported that the incorporation of insolation materials in the building can reduce the indoor temperature fluctuation up to 4 °C that would save 10–30% of energy usage [1,2].

Although many investigations have been carried out on utilizing waste glass in the form of glass powder [7–11] and glass bead [12–15] in concrete products, it has not found its position in the construction industry yet. Expanded glass aggregates (EGA), is a new commercial product, which are manufactured from waste and post-consumer glass. The EGA possess a relatively smooth surface with numerous encapsulated pores can be used as an insulating material [16]. The porous structure and low thermal conductivity of EGA can effectively reduce the heat transfer rate. The utilization of EGA in cementitious materials brings two-fold advantages, first, reducing the landfill cost and environment and secondly can reduce the energy consumption in buildings [16]. Recently, some studies [17–21] have been conducted to investigate the effect of EGA on mechanical and thermal properties of concrete and cement mortar, however, the utilization of EGA as an insulating material is at the initial stage. The great advantage of EGA is the possibility of production in a variety of size. Such a variety of particle sizes allows the improvement of the homogeneity of mixture and consequently reduces the possibility of segregation of the mixture [22].

Yu et al. [23] and Spiesz et al. [24] developed a cement-based lightweight composite using five different size of EGA (range between 0.1 and 2.0 mm) and reported the density of 1280–1490 kg/m^3 and compressive strength of 23.3–30.2 MPa. Rumsys et al. [21] prepared cement mortar with two types of fine expanded aggregates (expanded glass and expanded clay) to compare their compressive strength and durability properties. In the mixes, they replaced the fine aggregates with expanded glass and expanded clay by the weight of the sand (8.5, 16.7, 33.3, 66.7, and 100 wt%). The obtained results revealed that in the mixtures with 100% EGA, the density decreased about 37% and the compressive strength after 28 days of curing dropped about 60%. The results also confirmed that EGA could be applied in the cementitious composites without limitation related to the alkali-silica reaction. In the experiment conducted by Namsone et al. [25], a foamed matrix was prepared using EGA and the mechanical, thermal and frost resistance properties were examined. They obtained the compressive of 4.7 and 5.7 MPa at the age of 7 and 28 days and the thermal conductivity of 0.152–0.108 W/m.K. Moreover, it was observed that reference samples had lower values of weight loss (g/m^2) after the freeze–thaw test comparing to compositions with EGA. They also characterized the microstructure of the prepared foam matrix using optical microscopy and observed that EGA were distributed uniformly over the cross-section without any processes of segregation.

Abd Elrahman et al. [16] fabricated EGA-cement mortar and reported crushing resistance of 1.9–2.9 N/mm^2 and water absorption of 13.6–15.8 wt% depending on the particle size. The results showed a compressive strength of about 6 MPa and thermal conductivity less than 0.14 W/m.K. In the study conducted by [26], the influence of the grain size and percentage of EGA content on physical and mechanical properties of the cement composite were investigated. They reported an average porosity of 45–67% and bulk density of 903–1078 kg/m^3 in specimens containing 100% EGA with the size of 2–4 mm. Moreover, the compressive strengths of 6.68–12.49 MPa obtained for EGA cement mortar. In another attempt, [27] investigated the possibility of using artificial neural networks to design the composition of cement composite containing EGA with the desired properties. They established the relation between the quantity of EGA and the porosity, bulk density, and compressive strength of a cement composite. Moreover, previous studies revealed that high glass content (above 50%) could considerably increase the water absorption of cementitious composites [28,29]. It can be concluded that incorporation of EGA in cement mortar can significantly reduce the mechanical properties such as compressive strength and water resistance of cement matrix. Hence, in order to compensate the reduction in mechanical strength and water absorption of cementitious composites integrated with EGA, nanofillers such as TiO$_2$ can be used. Previous researches have demonstrated that the addition of TiO$_2$ nanoparticles effectively enhanced the compressive strength and reduce the water absorption of cementitious composites [30–35]. Indeed, nTiO$_2$ fills the nanovoids in concrete, which leads to the increment of compressive strength up to 40% [34,36]. Moreover, TiO_2 accelerates the formation of C-S-H gel by increasing the amount of crystalline $Ca(OH)_2$ at the early age of hydration [32,37]. Ma et al. [32] reported 37% and 44% increase in tensile and flexural strength respectively for the samples containing TiO_2. In addition, the results indicated that the addition of TiO_2 could significantly refine the pores and shift them to the harmless pores. In the research conducted by Behfarnia et al. [31], it was observed that TiO_2 nanoparticles decreased the permeability of the cement matrix. In the research conducted by Khushwaha et al. [34] and Sorathiya et al. [33], the effect of various proportion of TiO_2 was studied. It was concluded that addition of TiO_2 up to 1% could significantly enhance the mechanical properties of concrete.

This research aims to develop a cement mortar with a lower heat transfer rate and insulating properties using substitution of a natural aggregate (NA) with EGA. In this study, the effect of incorporation of EGA and TiO_2 nanoparticles on workability, bulk density, water penetration, compressive strength, and heat transfer rate of the cement mortar were investigated. Infrared thermography (IRT) was used to measure the thermal insulation property of EGA cement mortar. The IRT technique has been utilized to evaluate the thermal energy storage performance of building materials in previous studies [38], however it has not been used for measuring the thermal insulating property of the EGA cement mortar. The conducted research is an additional step toward development of insulating building material and sustainable application of EGA in the construction industry.

2. Materials and Methods

2.1. Materials

The materials used in the study to fabricate cement mortar composite were ordinary Portland cement (OPC), natural aggregate (NA), recycled expanded glass aggregate (EGA), superplasticizer (SP), and nano titanium dioxide ($nTiO_2$). OPC from Boral Australia Co. and in accordance with AS3972 was used as a binder and Sikament NN was used as a superplasticizer (SP) in the mix, which meets all requirements as per AS1478.1 for high range water reducing admixture. EGA with a particle size of 0.25-4 mm from EGT Co. is shown in Figure 1. The specifications of EGA are compliant with EN and DIN standards. Figure 2 shows a SEM image of the utilized EGA in this study. Table 1 demonstrates the physical, mechanical, and thermal properties of the EGA.

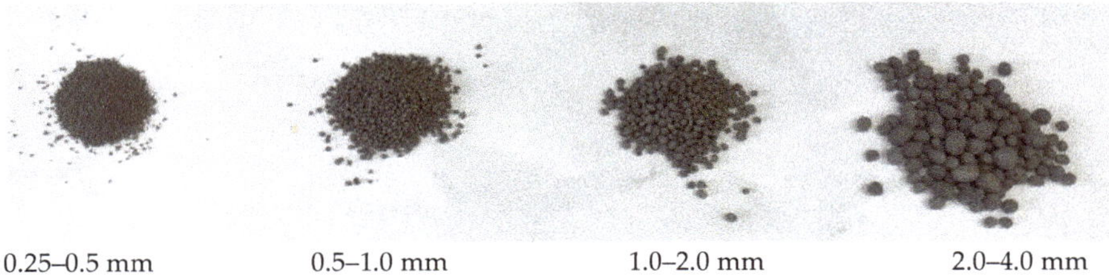

Figure 1. Expanded glass aggregate (EGA) with a different grain size.

Table 1. Physical, mechanical, and thermal properties of the EGA (Expanded Glass Technologies).

Property	Grain Size			
	0.25–0.5	**0.5–1**	**1–2**	**2–4**
Loose bulk density (kg/m^3)	300	250	220	190
Particle density (kg/m^3)	540	450	350	310
Compressive strength (MPa)	≥2.9	≥2.6	≥2.4	≥2.2
Thermal conductivity (W/mK)	0.07	0.07	0.07	0.07

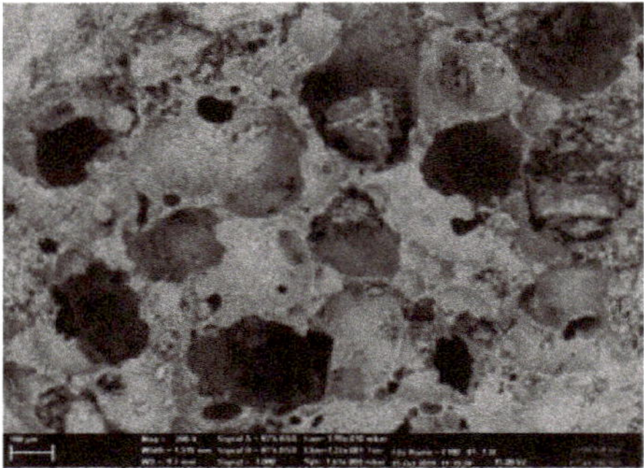

Figure 2. SEM pictures of the utilized EGA.

Crushed gravel with the maximum size of 4.0 mm and density of 2800 kg/m^3 was used as NA. The NA was subjected to the particle size distribution test to precisely replicate the distribution of NA for the replacement of EGA by volume in the cement mortar. The size distribution testing was completed in accordance with AS1012 and the results are found in Figure 3.

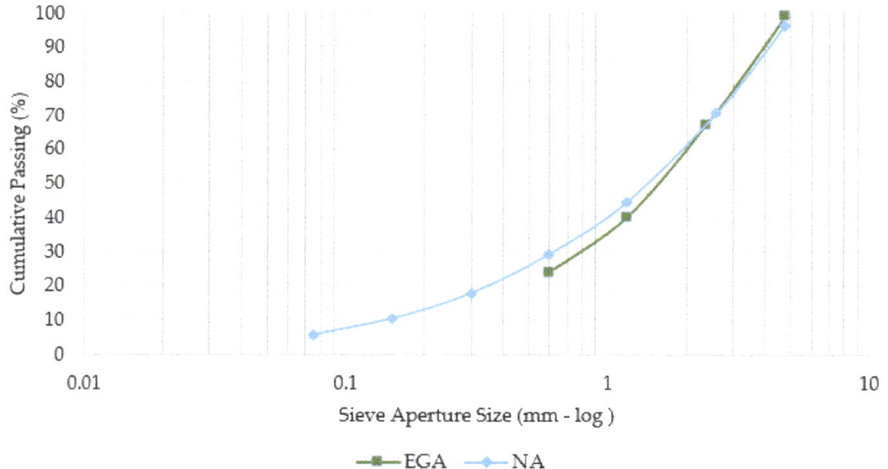

Figure 3. Size distribution of the natural aggregate and EGA.

Moreover, the mercury intrusion porosimetry (MIP) test was undertaken to measure the porosity as well as pore size distribution of the EGA. The MIP test results of the EGA are revealed in Figure 4.

Nanoparticles titanium dioxide (nTiO$_2$) purchased in the powder form from US Research Nanomaterials, Inc. Table 2 demonstrates the properties of the nTiO$_2$ as indicated by the manufacturer.

Table 2. The properties of the nano-nTiO$_2$ (US Research Nanomaterials, Inc.).

Properties	Value
Purity	99.98%
Average Particles Size	30 (nm)
Specific surface area	50 (m^2/g)
Bulk Density	0.42 (g/cm^3)
True Density	3.9 g/cm^3)
PH	5.5–6.5

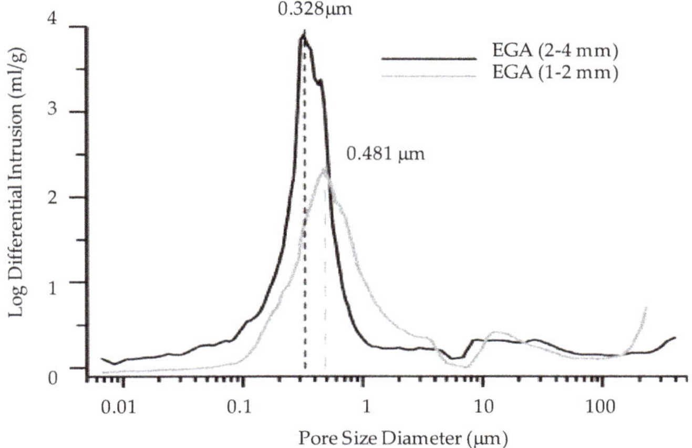

Figure 4. Pore size distribution of the EGA.

2.2. Sample Preparation

The mixes had a water to cement ratio of 0.4 and a sand to cement ratio of 3:1. Two set of mixes were prepared: the first set of mixes were fabricated by partial and full replacement of NA with EGA without inclusion of $nTiO_2$. The designed mixes with 0%, 50%, and 100% replacement percentage of EGA implied with CS, E50, and E100, respectively. The second set of mixes was fabricated by partial and full replacement of NA with EGA and incorporation of 1% $nTiO_2$. The designed mixes with incorporation of TiO_2 and the EGA replacement percentage of 0%, 50%, and 100% defined as CT, E50T, and E100T, respectively.

To fabricate the mixes, the dry materials (cement and NA/EGA) were placed in the mixer and mixed on the low speed for 2.0 min. In the case of CT, E50T, and E100T mixes, the $nTiO_2$ were sonicated for 15 min in the solution of water and superplasticizer (SP) [39]. Then the dispersed $nTiO_2$/SP/water solution was added slowly to the mix and the materials were mixed for another 5 min. The mixes cast in $70 \times 70 \times 70$ mm^3 cubes and demolded after 24 h. The samples were cured in the fog room at a constant temperature of 23 °C and in accordance with AS1012.8. Table 3 demonstrates the mix proportion of the samples. The abbreviations for labeling each mix are defined in a way that the letters C and E representing control sample and mortar sample containing EGA respectively and number after the letters presents the percentage of NA replacement with EGA into the mixture. The letter T demonstrates the presence of TiO_2 in the mix. For instance, the E50T mixture represents the sample that contains 50% EGA and TiO_2.

Table 3. Mix proportion of the samples (kg/m^3) of mortar.

Composite ID	NA	EGA	Cement	Water	S.P	$nTiO_2$
CS	1750	0	525	233	11.7	-
CT	1750	0	525	233	11.7	1%
E50	875	133	525	233	8.8	-
E50T	875	133	525	233	8.8	1%
E100	0	267	525	233	5.8	-
E100T	0	267	525	233	5.8	1%

2.3. Experimental Tests

The flow table test was undertaken on the fresh cement mortar samples in accordance to the AS2701 to measure the mixtures workability and consistency. Moreover, the density of the mixture was determined via the density test according to AS2701. To measure the water penetration of the specimens, the water absorption test was conducted in accordance with AS1012.21 at the age of 28 days. The compressive test was undertaken on the cube specimens with the size of 70 mm × 70 mm × 70 mm and in accordance with AS1012.9 at the age of 7, 14, and 28 days. For each test, three samples were tested and the average including the error bar were reported.

In this study, the thermal insulation property and heat transfer rate of cement mortar containing EGA was evaluated by measuring the surface temperature distribution using infrared thermal imaging camera. For this purpose, the specimens with the dimension of 70 mm × 70 mm × 30 mm were prepared and kept at about 27 °C for a few hours to allow all samples to achieve the same initial temperature. Then the samples were exposed to a heat source and the surface temperature distribution of the other side was captured by an infrared thermal camera for 15 min (Testo 872, Testo Australia). The thermal test was repeated for three times for each sample. Figure 5 illustrates a schematic diagram of the thermal test.

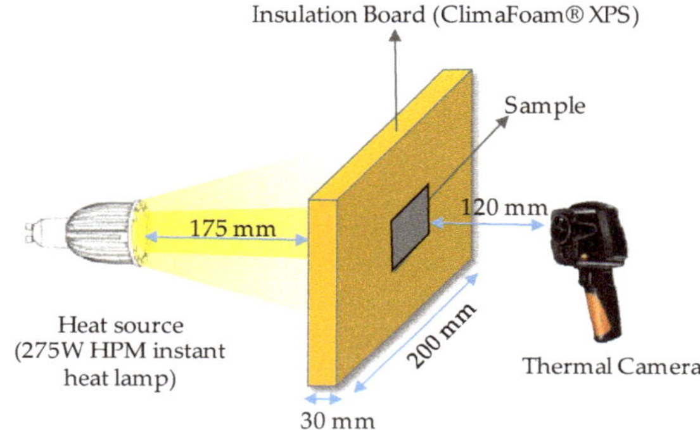

Figure 5. Infrared thermography test.

3. Results and Discussion

3.1. Workability

Figure 6 shows the flow table test and the flow table test results are presented in Table 4. The flow table test values were determined by averaging the diameters of each mixes test. All mixes showed flow values in the range of 140–215 mm, without segregation or bleeding. The results revealed that addition of EGA increased the workability of cement mortar up to 26.6% and 41.25% for the E50 and E100 mixes respectively compared to the control mix (CS). The increment trend in workability despite the decreasing on the amount of superplasticizer in E50 and E100 mixes is contributed to the smooth surface and spherical shape of EGA [40–42]. Adding to this, the increase in the flow values can be due to the increase in the amount of entrapped air voids. Furthermore, the workability of CT, E50T, and E100T mixes increased by 5.35%, 30.4%, and 53.7% respectively in comparison to the CS, E50, and E100 mixes respectively, which is attributed to the induction of the microbubble in the water solution during the sonication process and consequently increased in small air voids in the mixes.

Table 4. Flow results of mixes.

Mix ID	Average Flow Diameter (mm)
CS	140.0
E50	177.3
E100	201.3
CT	147.5
E50T	182.5
E100T	215.3

Figure 6. Flow table test.

3.2. Density

The density of the samples was measured, and the results are demonstrated in Figure 7. The measurement revealed the density of 2354, 1769, and 987 kg/m^3 for C.S, E50, and E100 respectively. It shows the density of E50 and E100 decreased 30% and 65% respectively in comparison to the CS, which is attributed to the very low density of EGA and its porous structure. In addition, the densities of CT, E50T, and E100T were 2%, 3%, and 6% higher than CS, E50, and E100 respectively. It can be concluded that the increase in density was attributed to the lower porosity in the cement matrix due to the incorporation of nTiO$_2$. It is noteworthy that E100 with density of 987 kg/m^3 was classified as a lightweight mortar that can be used for production of lightweight concrete. Figure 8 illustrates the cross section of CS and E100 samples.

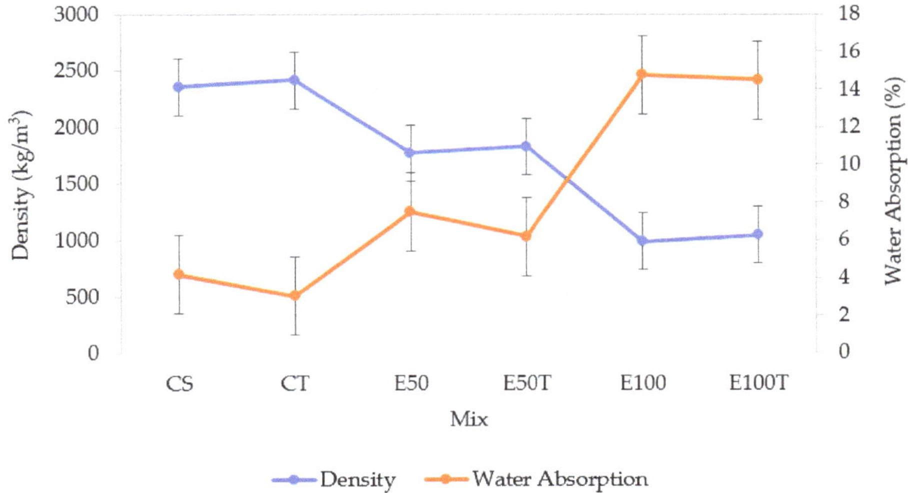

Figure 7. Density and water absorption of mix specimens.

3.3. Water Absorption

The water absorption test was completed on all mixes and the results are shown in Figure 7. The water absorption of 4.19% obtained for control sample (CS) however a higher water absorption rate was obtained for the mixes containing EGA. The water absorption of E50 and E100 mixes were 7.47% and 14.74% respectively, which shows a 78% and 252% increase in the permeability of the matrix, compared to the control sample. The increase in water absorption is due to the high porosity of EGA in comparison to NA. The results revealed that the water penetration increased by increasing the EGA content. The water absorption rate obtained for the E100 (values of approximately 14%) was higher than the acceptable range (<10%) [43,44]. The addition of $nTiO_2$ reduced the water absorption value by 28%, 17%, and 2% for samples containing 0%, 50%, and 100% EGA respectively. The decrease in water absorption upon the inclusion of $nTiO_2$ coincides with previous studies [45] and aligns with the density results. The reduction in water absorption was attributed to the filling effect of $nTiO_2$ and reducing the porosity of the cement matrix. It is worthy to note that the sonication process resulted in tiny bubbles of air uniformly distributed in the mortar. These small bubbles are like entraining air that improves the workability of the mixes. Indeed, $nTiO_2$ acted as nanofillers in mortar and improved the resistance to water penetration of the cement composite [46].

Figure 8. Cross-section of (**a**) the control sample (CS) and (**b**) E100 mixtures samples with uniform distribution of EGA in the cement matrix.

3.4. Compressive Strength

The experimental test for compressive strength was carried out at different curing ages of 7, 14, and 28 days. Figure 9 shows the impact of EGA and TiO_2 inclusion on the compressive strength of the mortar composites at different ages. It is observed that the inclusion of the EGA significantly decreased the compressive strength of cement mortar. The results of 28-day compressive strength demonstrated that 50% replacement of NA with EGA reduced the strength about 65.8% in compare to the control sample (C.S). In addition, it was observed that as the EGA content increased from 50% to 100%, the compressive strength dropped dramatically from 26.25 to 8.20 MPa at the age of 28 days. It is noteworthy that the compressive strength was still in the acceptable range and similar or higher than reported results in the literature [21,25]. Namsone et al. [25] reported the 28-day compressive of 5.7 MPa for a foamed matrix using EGA and obtained the compressive strengths of 6.68-12.49 MPa for the EGA cement mortar. Indeed, the samples containing 100% EGA without $nTiO_2$ had the lowest compressive strength out of all the mixes.

Furthermore, the results indicated a normal increasing trend for the compressive strength for CS, E50, and E100 mixes as the curing process progresses. However, the mixes containing $nTiO_2$ revealed a relatively different strength development tend. It was revealed that CT, E50T, and E100T mixes reached to 84.6%, 87.2%, and 77.2% of maximum strength within 7 days of curing while for samples without $nTiO_2$ (CS, E50, and E100 mixes) it happened at 14 days of curing. This behavior was attributed to the addition of $nTiO_2$ into the cementitious materials, which resulted in an accelerated rate of hydration process. A similar attribute has been reported in previous studies that when $nTiO_2$ is uniformly distributed throughout the matrix, the hydration process and formation of C-S-H gel is accelerated, which results in early strength [32,47,48]. In the other set of mixes, the effect of $nTiO_2$ inclusion on the compressive strength of mixes was investigated after a different curing time. The compressive strength results of E50T and E100T mixes at 28 days showed the similar trend. It was observed that the addition of EGA significantly decreased the compressive strength and the strength significantly dropped as the EGA content increased however, inclusion of $nTiO_2$ compensated some part of the compressive strength. The average compressive strength at 28 days of CT, E50T, and E100T mixes were 76.72, 29.70, and 11.4 MPa respectively, which shows 1.7%, 13.1%, and 39.0% enhancement in comparison to CS, E50, and E100 mixes respectively. It can be concluded that $nTiO_2$ acts as nanofillers in specimens and recovers their pore structure by decreasing voids and pores in the composite matrix [46].

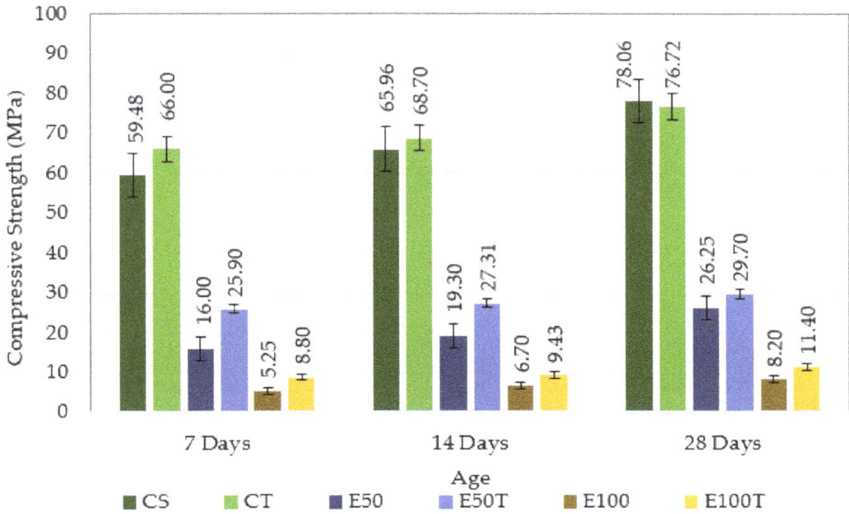

Figure 9. Compressive strength of samples at different ages.

In summary it can be concluded that the compressive strength and water absorption of concrete are highly influenced by the density of the mix. The results revealed an interrelationship between density and compressive strength. It was observed that the compressive strength dropped by decrement of the sample's density (CS, E50, and E100 mixes). Similarly, an increase in density for CT, E50T, and E100T mixes resulted in an increase in compressive strength compared to CS, E50, and E100 mixes respectively. Moreover, the results demonstrated an inverse relation between density and water absorption. It was found that water absorption increased by decreasing the density of the mixes in case of CS, E50, and E100. However, the water absorption decreased by integration of TiO2 into the mixes (CT, E50T, and E100T) and increment of density due to a lower porosity of the matrix.

3.5. Infrared Thermography

In order to evaluate the thermal insulating properties of the cement composites, the infrared thermography (IRT) experiment was carried out on all the samples: CS, E50, E100, CST, E50T, and E100T. Figure 10 illustrates the thermal images of surface temperature distribution of the samples captured by the IRT camera at different heating times. According to the relationship between the color and temperature value, it can be suggested that the heat-transferring rate and thermal conductivity of cement composites were significantly decreased with the inclusion of EGA. The thermal images clearly demonstrate a different temperature distribution in the control sample (CS mix) and the samples containing EGA (E50 and E100 mixes). The results show that the temperature increased rapidly in the CS however, a noticeable slower heat transfer rate observed for samples incorporated with EGA (E50 and E100). The data also revealed a drop in the heat transfer rate as the EGA content increased. For instance, after 15 min the average surface temperature in the CS sample reached 55 °C while the average surface temperature in the E50 and E100 samples reached 52.7 and 48.7 °C respectively, which shows a temperature difference of 2.3 °C and 6.0 °C for E50 and E100 respectively. Moreover, the results demonstrated the heat transfer rate of 1.75, 1.60, and 1.35 °C/min for CS, E50, and E100 respectively that shows a lower rate for samples containing EGA (E50 and E100) in comparison to the control sample (CS). This observation was attributed to a high porosity and low thermal conductivity of EGA. Indeed, by incorporation of EGA the air void is replaced with sand, which has a high thermal conductivity. EGA has a thermal conductivity of 0.07 W/mK that is much less than that of sand (Expanded Glass Technologies). Consequently, by replacing the NA with EGA the heat transfer of the cement composite was reduced.

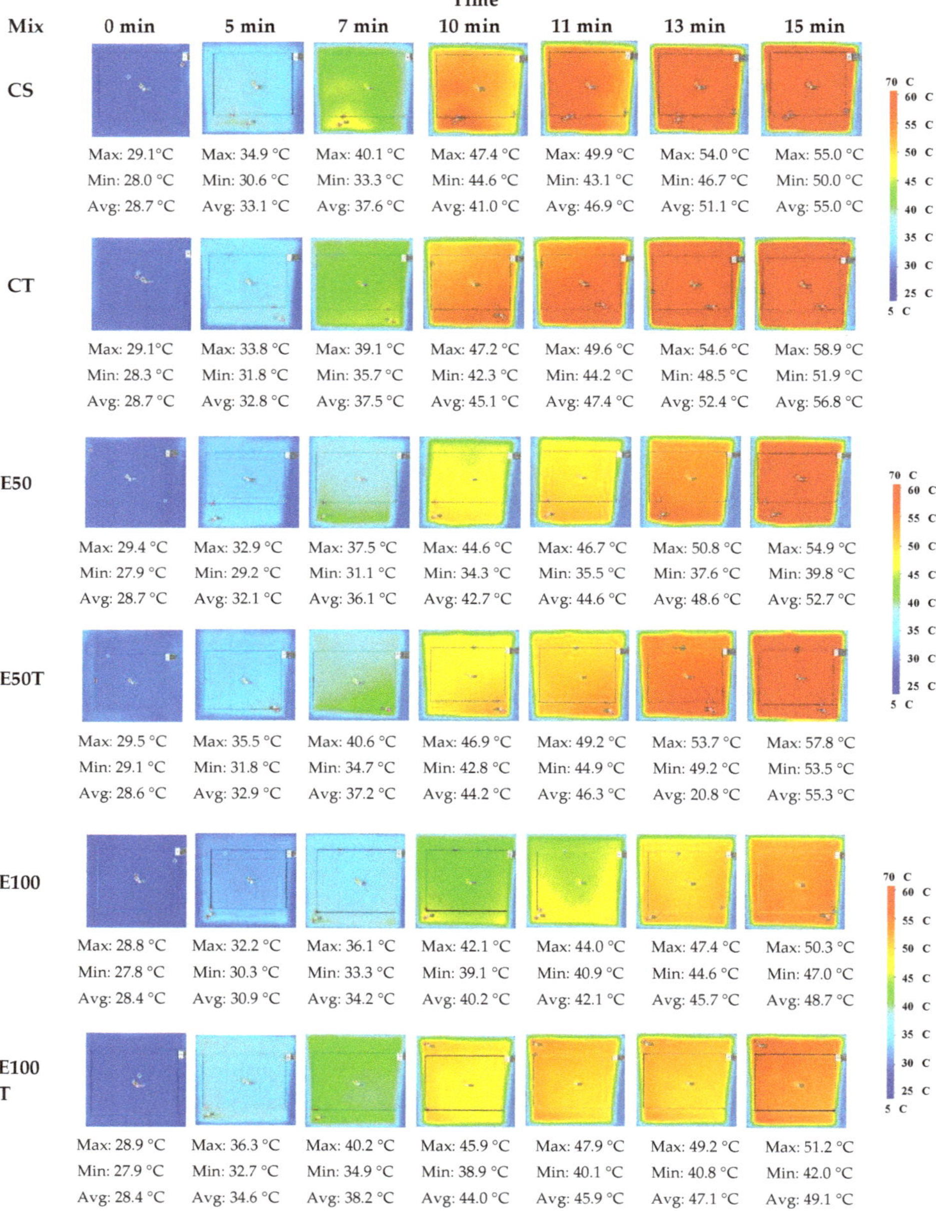

Figure 10. Infrared thermography images of different samples.

Table 5 demonstrates the average temperature differences for each mix. The thermal charging results for the samples inclusion $nTiO_2$ showed a different trend to the first set of mixes (mixes without $nTiO_2$). It was observed that incorporation $nTiO_2$ increased the heat transfer rate, which is undesired in terms of thermal insulation properties. The thermal images demonstrated that inclusion of $nTiO_2$ into the composite increased the heat transfer rate compared to the samples without $nTiO_2$. For example, after 15 min the average surface temperature in CS, E50, and E100 samples reached to 55 °C, 52.7 °C, and 48.7 °C respectively. While average surface temperature in the CT, E50T, and E100T samples reached to 56.8 °C, 55 °C, and 49.1 °C respectively that shows an increase in the temperature difference of 1.6 °C, 2.3 °C, and 1.4 °C respectively. Furthermore, the results demonstrated the heat transfer rate of 1.87, 1.76, and 1.38 °C/min for CT, E50T, and E100T respectively that indicates higher rate than the samples without $nTiO_2$. It can be concluded that $nTiO_2$ acts as a filler and changes the pore structures of the cement composite and consequently the thermal charging performance of the matrix. Therefore, in terms of thermal properties, NA substitution with EGA improves the thermal insulation properties of cement composites. This positive effect is attributed to lower thermally conductive and higher porosity of EGA compared to NA.

Table 5. The average temperature differences for each mix.

Mix	Δ ($T_{ave.}$)			Heat Transfer Rate (°C/min)		
	5 min	10 min	15 min	5 min	10 min	15 min
CS	4.40	12.30	26.30	1.58	2.80	1.75
CT	4.10	16.40	28.10	2.46	2.34	1.87
E50	3.40	14.00	24.00	2.12	2.00	1.60
E50T	4.30	15.60	26.40	2.26	2.16	1.76
E100	2.50	11.80	20.30	1.86	1.70	1.35
E100T	6.20	15.60	20.70	1.88	1.02	1.38

4. Conclusions

This experimental work investigated the physical properties as well as the thermal insulation property of cement mortar containing EGA and TiO_2. The findings revealed that incorporating EGA into the mortar composite causing a significant decrease in density and compressive strength, which was attributed to the porous nature and low compressive strength of EGA. The results also demonstrated that as the EGA content increased, the workability and water absorption of cement composite increased. It is found that the increase in water absorption was due to the high porosity of EGA in comparison to NA. However, the beneficial effect of the EGA was the decrease in the heat-transferring rate of the cement composite, which indicates the feasibility of a potential reduction in energy consumption in buildings. Moreover, the results demonstrated that inclusion of TiO_2 into the cement composite partially compensated the water absorption and loss in compressive strength. However, it was revealed that addition of $nTiO_2$ into EGA-cement composites increased the heat transfer rate of the cement matrix and insulation properties as $nTiO_2$ acts as nanofillers and changes the pores structure in the cement matrix. It can be concluded that in terms of thermal behavior, substitution of NA with EGA decreases the heat transfer rate and consequently improves the thermal insulation properties of the cement mortar.

Author Contributions: Conceptualization, A.Y., W.T.; methodology, W.T., M.K., S.W. and C.F.; Investigation, W.T., M.K. and A.Y.; writing—original draft preparation, A.Y., M.K.; writing—review and editing, W.T., C.F., M.K., S.W. and A.Y. All authors have read and agree to the published version of the manuscript.

Funding: This research received no external funding.

Acknowledgments: The authors gratefully acknowledge the financial support provided by University of Newcastle (2017 UNIPRS and UNRS Central Scholarship). We also thank the support and cooperation of the staff of the concrete laboratory at the School of Civil Engineering, University of Newcastle, and Expanded Glass Technologies Pty Ltd. for providing EGA for this experimental investigation.

Conflicts of Interest: The author declares no conflict of interest.

References

1.	Chung, O.; Jeong, S.G.; Kim, S. Preparation of energy efficient paraffinic PCMs/expanded vermiculite and perlite composites for energy saving in buildings. *Sol. Energy Mater. Sol. Cells* **2015**, *137*, 107–112. [CrossRef]
2.	Rao, V.V.; Parameshwaran, R.; Ram, V.V. PCM-mortar based construction materials for energy efficient buildings: A review on research trends. *Energy Build.* **2018**, *158*, 95–122. [CrossRef]
3.	Krishnamoorthy, R.R.; Zujip, J.A. Thermal conductivity and microstructure of concrete using recycle glass as a fine aggregate replacement. *Int. J. Emerg. Technol. Adv. Eng.* **2013**, *3*, 463–471.
4.	Real, S.; Gomes, M.G.; Moret Rodrigues, A.; Bogas, J.A. Contribution of structural lightweight aggregate concrete to the reduction of thermal bridging effect in buildings. *Constr. Build. Mater.* **2016**, *121*, 460–470. [CrossRef]
5.	Roberz, F.; Loonen, R.C.G.M.; Hoes, P.; Hensen, J.L.M. Ultra-lightweight concrete: Energy and comfort performance evaluation in relation to buildings with low and high thermal mass. *Energy Build.* **2017**, *138*, 432–442. [CrossRef]
6.	Sibahy, A.A.; Edwards, R. Thermal behaviour of novel lightweigh concrete at ambient and elevated temperatures: Experimental, modelling and parametric studies. *Constr. Build. Mater.* **2012**, *31*, 174–187. [CrossRef]
7.	Al-Zubaid, A.B.; Shabeeb, K.M.; Ali, A.I. Study the effect of recycled glass on the mechanical properties of green concrete. *Energy Procedia* **2017**, *119*, 680–692. [CrossRef]
8.	Park, S.B.; Lee, B.C.; Kim, J.H. Studies on mechanical properties of concrete containing waste glass aggregate. *Cem. Concr. Res.* **2004**, *34*, 2181–2189. [CrossRef]
9.	Islam, G.M.S.; Rahman, M.H.; Kazi, N. Waste glass powder as partial replacement of cement for sustainable concrete practice. *Int. J. Sustain. Built Environ.* **2017**, *6*, 37–44. [CrossRef]
10.	Aliabdo, A.A.; Abd Elmoaty, A.E.M.; Aboshama, A.Y. Utilization of waste glass powder in the production of cement and concrete. *Constr. Build. Mater.* **2016**, *124*, 866–877. [CrossRef]
11.	Afshinnia, K.; Rangaraju, P.R. Efficiency of ternary blends containing fine glass powder in mitigating alkali–silica reaction. *Constr. Build. Mater.* **2015**, *100*, 234–245. [CrossRef]
12.	Yun, T.S.; Jeong, Y.J.; Han, T.S.; Youm, K.S. Evaluation of thermal conductivity for thermally insulated concretes. *Energy Build.* **2013**, *61*, 125–132. [CrossRef]
13.	Chung, S.Y.; Abd Elrahman, M.; Sikora, P.; Rucinska, T.; Horszczaruk, E.; Stephan, D. Evaluation of the effects of crushed and expanded waste glass aggregates on the material properties of lightweight concrete using image-based approaches. *Materials* **2017**, *10*, 1354. [CrossRef]
14.	Chung, S.Y.; Han, T.S.; Kim, S.Y.; Kim, J.H.J.; Youm, K.S.; Lim, J.H. Evaluation of effect of glass beads on thermal conductivity of insulating concrete using micro CT images and probability functions. *Cem. Concr. Compos.* **2016**, *65*, 150–162. [CrossRef]
15.	Ez-zaki, H.; El Gharbi, B.; Diouri, A. Development of eco-friendly mortars incorporating glass and shell powders. *Constr. Build. Mater.* **2018**, *159*, 198–204. [CrossRef]
16.	Abd Elrahman, M.; Chung, S.Y.; Stephan, D. Effect of different expanded aggregates on the properties of lightweight concrete. *Mag. Concr. Res.* **2018**, *71*, 95–107. [CrossRef]
17.	Yu, Q.L.; Spiesz, P.; Brouwers, H.J.H. Ultra-lightweight concrete: Conceptual design and performance evaluation. *Cem. Concr. Compos.* **2015**, *61*, 18–28. [CrossRef]
18.	Yu, R.; van Onna, D.V.; Spiesz, P.; Yu, Q.L.; Brouwers, H.J.H. Development of ultra-lightweight fibre reinforced concrete applying expanded waste glass. *J. Clean. Prod.* **2016**, *112*, 690–701. [CrossRef]
19.	Šeputytė-Jucikė, J.; Sinica, M. The effect of expanded glass and polystyrene waste on the properties of lightweight aggregate concrete. *Eng. Struct. Technol.* **2016**, *8*, 31–40. [CrossRef]
20.	Maddalena, C.; Luca, B. Durability of lightweight concrete with expanded glass and silica fume. *Mater. J.* **2017**, *114*. [CrossRef]
21.	Rumsys, D.; Spudulis, E.; Bacinskas, D.; Kaklauskas, G. Compressive strength and durability properties of structural lightweight concrete with fine expanded glass and/or clay aggregates. *Materials* **2018**, *11*, 2434. [CrossRef] [PubMed]

22. Bogas, J.A.; Gomes, A.; Pereira, M. Self-compacting lightweight concrete produced with expanded clay aggregate. *Constr. Build. Mater.* **2012**, *35*, 1013–1022. [CrossRef]

23. Yu, Q.; Spiesz, P.; Brouwers, H. Development of cement-based lightweight composites–Part 1: Mix design methodology and hardened properties. *Cem. Concr. Compos.* **2013**, *44*, 17–29. [CrossRef]

24. Spiesz, P.; Yu, Q.; Brouwers, H. Development of cement-based lightweight composites–Part 2: Durability-related properties. *Cem. Concr. Compos.* **2013**, *44*, 30–40. [CrossRef]

25. Namsone, E.; Sahmenko, G.; Namsone, E.; Korjakins, A. Thermal conductivity and frost resistance of foamed concrete with porous aggregate, Environment. Technology. Resources. In Proceedings of the 11th International Scientific and Practical Conference, Rezekne, Latvia, 15–17 June 2017; Volume 3, pp. 222–228.

26. Kurpińska, M.; Ferenc, T. Effect of porosity on physical properties of lightweight cement composite with foamed glass aggregate. In Proceedings of the II International Conference of Computational Methods in Engineering Science (CMES'2017), Lublin, Poland, 23–25 November 2017; p. 06005.

27. Kurpinska, M.; Kułak, L. Predicting performance of lightweight concrete with granulated expanded Glass and Ash aggregate by means of using Artificial Neural Networks. *Materials* **2019**, *12*, 2002. [CrossRef]

28. Lee, G.; Poon, C.S.; Wong, Y.L.; Ling, T.C. Effects of recycled fine glass aggregates on the properties of dry–mixed concrete blocks. *Constr. Build. Mater.* **2013**, *38*, 638–643. [CrossRef]

29. Penacho, P.; Brito, J.D.; Rosário Veiga, M. Physico-mechanical and performance characterization of mortars incorporating fine glass waste aggregate. *Cem. Concr. Compos.* **2014**, *50*, 47–59. [CrossRef]

30. Lucas, S.S.; Ferreira, V.M.; de Aguiar, J.L.B. Incorporation of titanium dioxide nanoparticles in mortars—Influence of microstructure in the hardened state properties and photocatalytic activity. *Cem. Concr. Res.* **2013**, *43*, 112–120. [CrossRef]

31. Behfarnia, K.; Azarkeivan, A.; Keivan, A. The effects of TiO_2 and ZnO nanoparticles on physical and mechanical properties of normal concrete. *Asian J. Civ. Eng. (Bhrc)* **2013**, *14*, 517–531.

32. Ma, B.; Li, H.; Mei, J.; Li, X.; Chen, F. Effects of nano-TiO_2 on the toughness and durability of cement-based material. *Adv. Mater. Sci. Eng.* **2015**, *2015*, 583106. [CrossRef]

33. Sorathiya, J.; Shah, S.; Kacha, M.S. Effect on addition of nano "titanium dioxide"(TiO_2) on compressive strength of cementitious concrete. *Kalpa Publ. Civ. Eng.* **2017**, *1*, 219–225.

34. Khushwaha, A.; Saxena, R.; Pal, S. Effect of titanium dioxide on the compressive strength of concrete. *J. Civ. Eng. Environ. Technol.* **2015**, *2*, 482–486.

35. Grebenisan, E.; Hegyi, A.; Szilágyi, H. A review on developing self-cleaning cementitious materials. *Constructii* **2018**, *19*, 37–43.

36. Nazari, A.; Riahi, S.; Riahi, S.; Shamekhi, S.F.; Khademno, A. Assessment of the effects of the cement paste composite in presence TiO_2 nanoparticles. *J. Am. Sci.* **2010**, *6*, 43–46.

37. Nazari, A.; Riahi, S. TiO_2 nanoparticles' effects on properties of concrete using ground granulated blast furnace slag as binder. *Sci. China Technol. Sci.* **2011**, *54*, 3109. [CrossRef]

38. Cui, H.; Feng, T.; Yang, H.; Bao, X.; Tang, W.; Fu, J. Experimental study of carbon fiber reinforced alkali-activated slag composites with micro-encapsulated PCM for energy storage. *Constr. Build. Mater.* **2018**, *161*, 442–451. [CrossRef]

39. Yousefi, A.; Bunnori, N.M.; Khavarian, M.; Hassanshahi, O.; Majid, T.A. Experimental investigation on effect of multi-walled carbon nanotubes concentration on flexural properties and microstructure of cement mortar composite. In Proceedings of the International Conference of Global Network for Innovative Technology (IGNITE) and AWAM International Conference in Civil Engineering (AICCE) 2017, Penang, Malaysia, 8–10 August 2017; AIP Conference Proceedings. AIP Publishing: Melville, NY, USA, 2017; p. 020032. [CrossRef]

40. Ling, T.C.; Poon, C.S.; Kou, S.C. Influence of recycled glass content and curing conditions on the properties of self-compacting concrete after exposure to elevated temperatures. *Cem. Concr. Compos.* **2012**, *34*, 265–272. [CrossRef]

41. Sharifi, Y.; Houshiar, M.; Aghebati, B. Recycled glass replacement as fine aggregate in self-compacting concrete. *Front. Struct. Civ. Eng.* **2013**, *7*, 419–428. [CrossRef]

42. Wright, J.R.; Cartwright, C.; Fura, D.; Rajabipour, F. Fresh and hardened properties of concrete incorporating recycled glass as 100% sand replacement. *J. Mater. Civ. Eng.* **2013**, *26*, 04014073. [CrossRef]

43. Medina, C.; Zhu, W.; Howind, T.; de Rojas, M.I.S.; Frías, M. Influence of mixed recycled aggregate on the physical–mechanical properties of recycled concrete. *J. Clean. Prod.* **2014**, *68*, 216–225. [CrossRef]

44. Borhan, M.; Mohamed Sutan, N. Laboratory study of water absorption of modified mortar. *Unimas E J. Civ. Eng.* **2011**, *2*, 25–30.
45. Qudoos, A.; Kim, H.; Ryou, J.S. Influence of titanium dioxide nanoparticles on the sulfate attack upon ordinary Portland cement and slag-blended mortars. *Materials* **2018**, *11*, 356.
46. Nazari, A.; Riahi, S. The effect of TiO_2 nanoparticles on water permeability and thermal and mechanical properties of high strength self-compacting concrete. *Mater. Sci. Eng. A* **2010**, *528*, 756–763. [CrossRef]
47. Nehdi, M.; Soliman, A.M. Early-age properties of concrete: Overview of fundamental concepts and state-of-the-art research. *Proc. Inst. Civ. Eng. Constr. Mater.* **2011**, *164*, 57–77. [CrossRef]
48. Lee, B.Y.; Jayapalan, A.R.; Kurtis, K.E. Effects of nano-TiO_2 on properties of cement-based materials. *Mag. Concr. Res.* **2013**, *65*, 1293–1302. [CrossRef]

© 2020 by the authors. Licensee MDPI, Basel, Switzerland. This article is an open access article distributed under the terms and conditions of the Creative Commons Attribution (CC BY) license (http://creativecommons.org/licenses/by/4.0/).

Article

Experimental Study on the Seismic Performance of Recycled Concrete Hollow Block Masonry Walls

Chao Liu [1],*, Xiangyun Nong [1], Fengjian Zhang [1], Zonggang Quan [2] and Guoliang Bai [1]

[1] College of Science, Xi'an University of Architecture and Technology, Xi'an 710055, China;
    nongxiangyun@xauat.edu.cn (X.N.); zfj9608@163.com (F.Z.); guoliangbai@126.com (G.B.)
[2] Xi'an Research & Design Institute of Wall and Roof Materials, Xi'an 710055, China; quanzg01@163.com
* Correspondence: chaoliu@xauat.edu.cn; Tel.: +86-180-9256-1062

Received: 20 September 2019; Accepted: 13 October 2019; Published: 15 October 2019

Abstract: This paper aims to manufacture recycled concrete hollow block (RCHB) which can be used for the masonry structure with seismic requirements. Five RCHB masonry walls were tested under cyclic loading to evaluate the effect of the axial compression stress, aspect ratio, and the materials of structural columns on the seismic performance. Based on the test results, the failure pattern, hysteresis curves, lateral drift, ductility, stiffness degradation, and the energy dissipation of the specimens were analyzed in detail. The results showed that with the increase of aspect ratios, the ductility of RCHB masonry walls increased, but the horizontal bearing capacity and energy dissipation of RCHB masonry walls decreased. With the increase of compressive stress, the bearing capacity and energy dissipation performance of RCHB masonry walls were improved, and the stiffness degraded slowly. The results also demonstrated that the RCHB masonry walls with structural columns, depending on whether the structural columns were prepared by ordinary concrete or recycled concrete, could increase the bearing capacity, ductility, and energy dissipation of specimens. The research confirmed that RCHB masonry walls could meet the seismic requirements through thoughtful design. Therefore, this study provided a new cleaner production for the utilization of construction waste resources.

Keywords: construction and demolition wastes; resource utilization; recycled concrete hollow block; masonry walls; seismic performance

---

1. Introduction

Since the beginning of human civilization, construction and demolition wastes (CDW) had been becoming a global problem that affects the sustainable development of the resources and the environment [1]. To date, more than a hundred billion tons of CDW were generated in the world, while approximately 30% to 50% of them were waste concrete [2]. However, this considerable amount of waste concrete is mainly disposed of in landfills, resulting in a severe environmental problem [3–6]. In order to protect the ecological environment and achieve sustainable development, reducing, reusing, and recycling construction waste is a desirable approach to preserve the ecological environment [7–11]. Therefore, many countries have passed legislation to encourage the recycling of waste concrete for the resource utilization of recycled concrete.

Recycled concrete aggregates (RCA) can be obtained from waste concrete by crushing, screening, cleaning, and separating [12,13]. Previous studies have been conducted to characterize the potential advantages and drawbacks of RCA. Being different from natural aggregates (NA), RCA has a layer of old mortar attached to the surface, which has more loose pores at the interface [14]. The attached old mortar brings with it worse properties: a lower apparent density and higher water absorption [15,16]. Consequently, these properties cause the mechanical properties of RCA to be inferior to that of NA, which limits the application of recycled concrete in civil engineering. Although the mechanical properties of recycled aggregates are poor, different scholars, institutions, and countries have been applying recycled aggregates into structures through rational design and experiment in recent years, and in some cases, good results have been achieved [17–23].

To further promote the recycling of construction waste, growing studies on the production of recycled concrete blocks prepared by RCA have been conducted in years. Researchers proposed that the mechanical properties of recycled concrete blocks mainly depend on the substitution rate and qualities of RCA. Soutsos et al. [24] pointed out that the increase of recycled fine aggregates content has a more significant impact on the reduction of recycled concrete block strength. Therefore, the maximum replacement rate of recycled fine aggregates is recommended to be 20%. Bai et al. [25] also confirmed the findings through a similar experiment. A study conducted by Poon et al. [26] aimed to investigate the mechanical properties of recycled concrete blocks. They found that the substitution rate of RCA below 50% had little effect on the compressive strength of recycled concrete blocks. However, the compressive strength of recycled concrete hollow block (RCHB) decreases with the increase of the content of low-grade RCA (The content of soil or broken brick in aggregates being >10%) [27]. Guo et al. [28] investigated the influence of different substitution rates of recycled concrete aggregates on the mechanical properties of recycled concrete blocks. The test results illustrated that with the substitution rate up to 75%, the strength of recycled concrete blocks slightly decreases but still complies with the standards. An experiment by Sabai et al. [29] showed that the compressive strength of recycled concrete blocks with 100% RCA can achieve the target of 7 MPa, and even the minimum strength requirement of construction.

Despite the fact that the compressive strength of the concrete block is the critical performance index about whether the block can be used in a masonry structure, the mechanical performance of a masonry prism is more reflective of the actual stress state of the masonry structure. Corinaldesi et al. [30] studied the compressive, shear, and bond strength of recycled mortar prisms, and found that the shear and compressive strength of recycled mortar prisms were close to or even better than that of ordinary mortar prisms. Guo et al. [28] conducted a study of recycled concrete block prisms, and the conclusions are consistent with that of Corinaldesi et al. [30]. In general, the shear strength of the recycled concrete prisms is close to that of the ordinary concrete prisms. However, the seismic performance of the masonry structure, especially the hysteretic characteristics and energy dissipation capacities of the masonry is unclear by the tests of the masonry prisms.

Through the analysis of the above examples, it can be proved that the mechanical properties of recycled blocks can satisfy the requirement of practical application, and the shear properties of masonry assemblages fabricated by recycled concrete blocks are close to those of ordinary masonry assemblages. However, there is still a lack of knowledge about the recycled concrete blocks, and whether they can be used to produce masonry structures with seismic requirements, especially regarding the research about the seismic performance of masonry structures, which is crucial for addressing the utilization problem of RCA. Therefore, if the waste concrete can be recycled to produce RCHB which can be used in the structures with seismic requirements, this type of RCHB will be popularized and further applied on a larger scale.

In view of this, RCA generated from waste concrete was used to produce a new type of RCHB which can be used for masonry structures. In this experiment, three RCHB masonry walls without structural columns constraint, one RCHB wall constrained by the recycled aggregate concrete structural columns, and other walls constrained by ordinary concrete structural columns were manufactured for seismic performance testing. The seismic behavior of the specimens, such as the failure pattern, the hysteresis curves, the skeleton curves, the ductility coefficient, and the energy dissipation of the specimens was analyzed under cyclic loading. The influences of aspect ratio, vertical axial stress, and different materials used for structural columns on the seismic performance of RCHB masonry walls were also studied. Finally, the seismic capacity of RCHB masonry structure under seismic loading was attained.

2. Experimental Program

2.1. Materials

Ordinary Portland cement with a 28 d strength grade of 42.5 MPa was employed in this experiment. Ordinary mortar with a standard 28 d strength grade of 10 MPa was used as the binder. The fine NA with a grain size of 0–5 mm were river sand. RCA was provided by Shaanxi Jianxin Technology Environmental Protection Co., Ltd. The RCA with a grain size of 5–10 mm (Figure 1) were crushed, cleaned, sieved, and separated from waste concrete. The fine NA and coarse NA were replaced by 0% recycled fine aggregates, and 100% RCA, respectively, and the mixture proportion of recycled aggregate concrete (RAC) is shown in Table 1. The grading of aggregates and properties of aggregates are shown in Figure 2 and Table 2, respectively. Based on the Chinese standard GB/T 41,112,013 [31] and the GB/T 8239-2014 [32], RCHB, prepared by RAC, with a compressive strength of 10 MPa were used for the fabrication of masonry walls, and the dimension of RCHB is 390 mm × 240 mm × 190 mm (Figure 3). A hot-rolled ribbed (HRB335) steel bar with a yielding strength of 335 MPa was used as longitudinal rebars. A hot-rolled plain (HPB235) steel bar with yielding strength of 235 MPa was selected as stirrups.

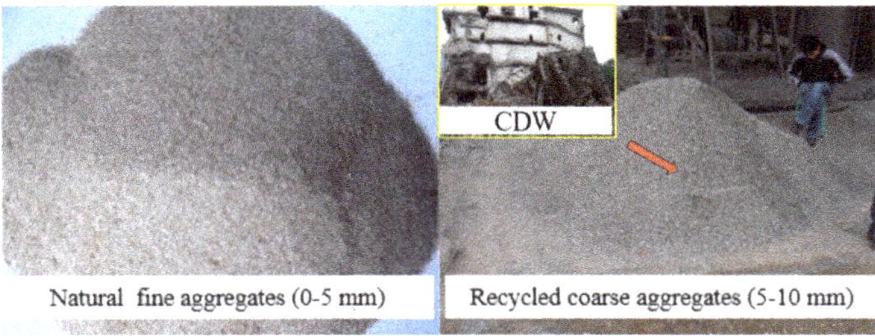

Figure 1. Aggregate components.

Table 1. Composition of recycled concrete block.

RCHB Components	Mixture Proportion/kg·m^{-3}
Water	150
Cement	375
Recycled coarse aggregates	945
Recycled fine aggregates	—
Natural coarse aggregates	—
Natural fine aggregates	630

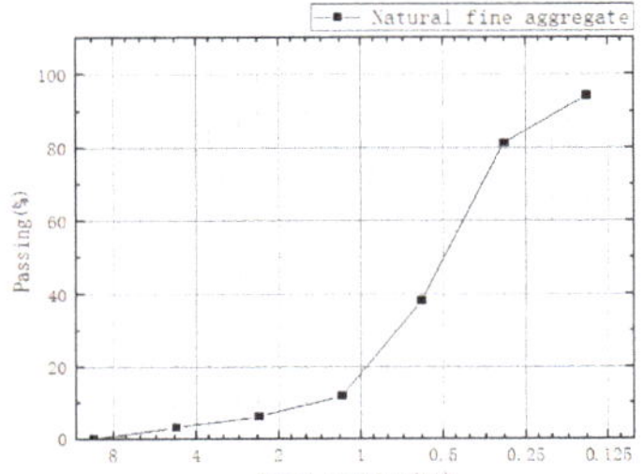

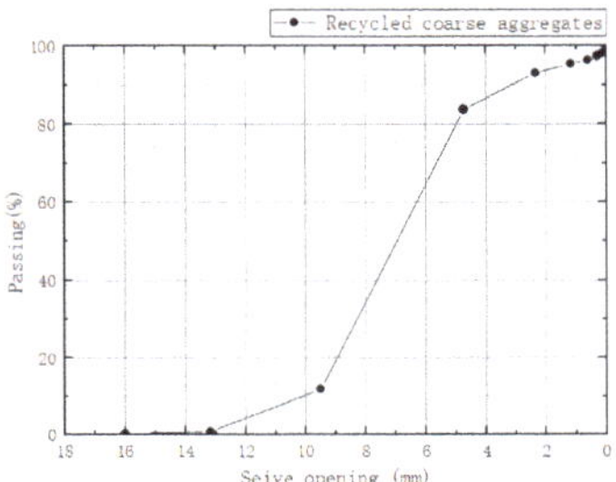

Figure 2. Fuller grading curves.

Table 2. Properties of aggregates.

Aggregate	Grading /mm	Apparent Density /kg·m^{-3}	Bulk Density /kg·m^{-3}	Water Absorption/%	Crushing Index	Fineness Modulus
NA	<5	2724.8	1448.4	1.02	—	2.18
RCA	5–10	2521.7	1096.1	3.50	24.74	—

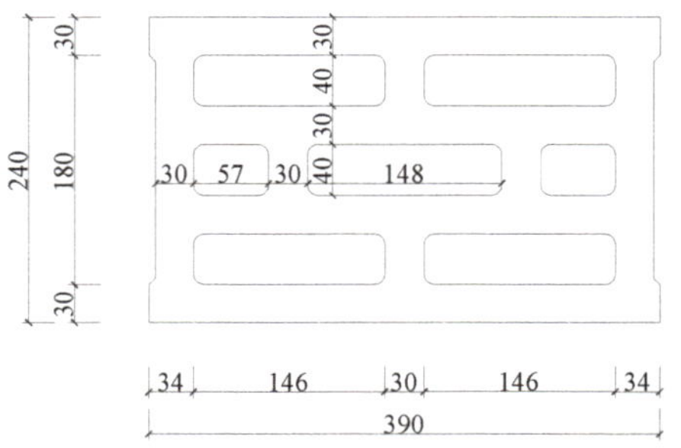

Figure 3. The dimension of recycled concrete hollow block.

2.2. Design of Specimens

Figure 4 shows the masonry procedure of all the specimens, including masonry the bottom beams, infill walls, structural columns, and ring beams. It is important to mention that ordinary mortar with a strength of 10 MPa was selected to masonry the mortar joints of the infill walls, and the fullness degree of mortar joints should be up to 80% to ensure the propagation of shear forces according to the seismic design requirements. In addition, the horizontal joint width was required to range from 8 to 12 mm.

Figure 4. Process of the construction of masonry wall specimens.

The experimental parameters used to determine the seismic performance of specimens were the compressive stress, aspect ratios, and the materials of the structural columns. The test specimens were divided into two groups. The first group consisted of three walls W1, W2, and W3, and the cyclic loading tests of them were conducted to analyze the influence of different compressive stress and aspect ratio on the seismic capacity. The second group contained two specimens, W4 and W5, which were constrained by structural columns, and the tests of them were conducted to analyze the influence of structural columns of different materials on the seismic capacity. The other components of all specimens, such as bottom beams and ring beams, were consistent in the test. Additionally, the value of the compressive stress was 0.6 MPa and 0.9 MPa, respectively, corresponding to the compressive stress of the upper weight on the middle floor and the bottom floor of an ordinary 9-storey residential building, respectively. Properties and dimensions of the five masonry walls are shown in Table 3 and Figure 5.

Table 3. Design of recycled concrete hollow block (RCHB) masonry walls.

No.	W1	W2	W3	W4	W5	Unit
Width	2000	2000	1200	2240	2240	mm
Height	1400	1400	2000	1350	1350	mm
Aspect ratio	0.700	0.700	1.667	0.603	0.603	—
Vertical load	288	432	259	323	323	kN
Vertical Axial stress	0.6	0.9	0.9	0.6	0.6	MPa
Materials of Structural columns	—	—	—	NA	RCA	—

Key: 2Φ10=2 bars of 335MPa

Φ6@200= 6mm stirrup bars of 300MPa with 200mm spacing

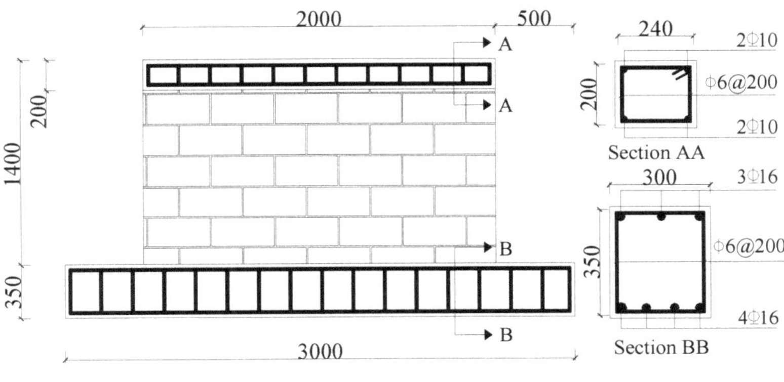

(**a**) W1 and W2.

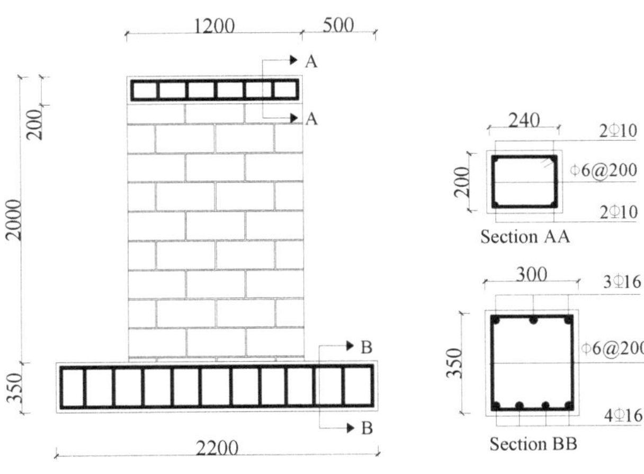

(**b**) W3

Figure 5. *Cont.*

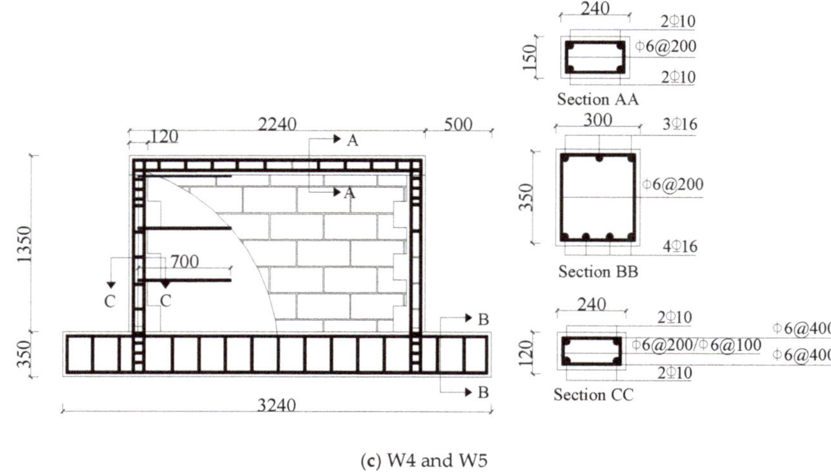

(**c**) W4 and W5

Figure 5. Size of the specimens: (**a**) W1 and W2; (**b**) W3; (**c**) W4 and W5.

The dimensions and the steel arrangement of the specimens should be meeting the specifications as outlined in GB/T 41112013. The section size of the structural column, ring beam, and bottom beam were 240 mm × 120 mm, 240 mm × 200 mm (or 240 mm × 150 mm) and 300 mm × 350 mm, respectively. Ordinary Portland concrete with 28 d axial strength of 42.5 MPa was used in the bottom beams during the test. The reinforcement arrangement is shown in Figure 5.

2.3. Test Procedure

The loading procedure included two steps: a load-controlled step and a drift-controlled step, which can be observed in Figure 6. First, the specimens were subjected to a vertical pre-compression load, which was kept constant during each test (Figure 2). After 20 min, 20 kN lateral loads were applied to the specimens in advance and repeated three times. Before formal loading, the experiment instruments were checked to see if they work properly, and then the experiment was conducted. During the load-controlled stage, the force control loading was adopted in the experiment. During this time, the horizontal load was applied cyclically with an increment of 40 kN until the walls cracked. After the walls initially cracked, the application method was changed to the drift control step. Then, the displacement levels of the cracked point were regarded as the first displacement step and set as the increment for the subsequent displacement cycles. Cyclic loading was applied twice at each displacement magnitude until wall failure. Subsequently, the tests were terminated when the lateral load of specimens decreased to about 85% of its peak value, or the number of specimen cracks reached about 50% of the total number of masonry mortar joints.

2.4. Test Device and Measuring Arrangement

The test device and the measuring instrument of deformation are presented in Figure 7. The layout of the measuring instruments and data collections were introduced as follows: (1) A horizontal linear variable differential transformer (LVDT) was placed in the middle of the ring beam to measure the horizontal displacement of the wall; (2) force sensors continuously recorded the values of vertical load and horizontal load applied by the vertical jack and the horizontal actuator; (3) the status of cracks was observed by the naked eye during experiments during the test. The occurrence, development, width of cracks, and maximum crack width were continually recorded at every loading process, and a marking pen described the shape of the cracks. The number and the values of cycles were also recorded; (4) two vertical LVDTs were placed on both sides of the bottom beam respectively to measure vertical displacement. A horizontal LVDT was arranged at one side of the bottom beams to measure horizontal displacement.

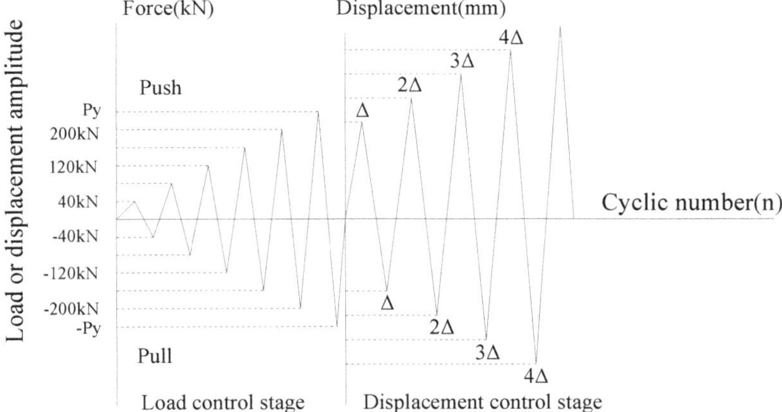

Figure 6. Lateral loading process.

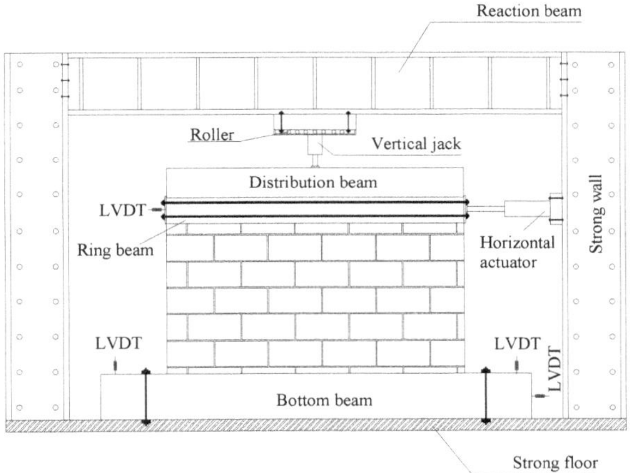

Figure 7. Schematic representation of the test setup.

3. Test Results

3.1. Failure Phenomenon and Pattern

The failure patterns of the five specimens are presented in Figure 8. All the specimens experienced three phases, such as cracking, peak, and failure in the experiment, but the failure process of the specimens W1, W2, and W3 were different from the specimens W4 and W5. Without structural columns, the specimens W1, W2, and W3 (bare walls) reached the ultimate strength and were destroyed immediately after the walls cracked, indicating the specimens had poor ductility and bearing capacity. On the contrary, the specimens W4 and W5 (reinforced walls) reflected better ductility and bearing capacity due to the constraints of the structural columns and ring beams. The analysis of the failure patterns of all the specimens is as follows:

1. In the initial stage of loading, horizontal cracks occurred on both sides of the wall root of the specimen W1 due to the flexural effect. By this time, no obvious cracks occurred in the diagonal

direction because the shear strength of the diagonal direction did not reach the cracked load. As the horizontal load increased, the horizontal cracks propagated rapidly and linked together gradually at the wall root, and the visible flexural effect was more apparent under horizontal loading. When the horizontal load reached the ultimate load, the mortar joint between the infill wall and the bottom beam was destroyed completely. Meanwhile, the specimen hardly bore the lateral load, indicating that specimen W1 failed. Therefore, the failure process of W1 was predominantly affected by the flexural effect;

2. During the initial loading stage, small horizontal cracks gradually appeared between the horizontal mortar cushion and the bottom beam of the specimen W2. As the horizontal load increased, the horizontal cracks of the wall root appeared along the ladder joints from the lower to the upper. When the lateral loads reached 80% of its ultimate loads, the cracks of the ladder joints continually widened and propagated simultaneously towards the diagonal direction. At this moment, numerous cracked blocks were spalled from the diagonal of the wall. When the lateral load increased to the ultimate load, the previously formed cracks at the diagonal direction linked together, and the bearing capacity of the specimen W2 declined rapidly, indicating that the specimen was destroyed;

3. Due to the large aspect ratio, the flexural influence on W3 was evident under the horizontal load, and the failure pattern of W3 was similar to W1;

4. In the initial loading stage, small stepped cracks appeared along the wall root and the vertical joints of the infill walls of the specimens W4 and W5. As the horizontal load increased, tiny horizontal cracks gradually appeared on both sides of the bottom of the structural columns, and other small cracks appeared at the upper oblique of the structural columns. As the horizontal load further increased, these cracks gradually widened and nearly connected, tearing apart the RCHB of the diagonal direction. When the lateral loads neared the ultimate loads, the diagonal cracks and horizontal cracks widened rapidly and linked together to form a large "X" shape crack under cyclic loading. At the same time, the spalling of the cracked blocks was observed in the diagonal direction of the specimens. Subsequently, the bearing capacity of the specimens reduced rapidly to 15% of its maximum values, indicating that the specimens W4 and W5 were destroyed.

3.2. Characteristic Load and Displacement

As can be noticed in Table 4, the ultimate load and the cracking load of the specimens W1 was 98% and 185% lower than that of W2, respectively, which showed that increasing the vertical load could increase the bearing capacity of specimens. The cracking load and the ultimate load of the specimen W2 was higher compared to W3, which indicated that the bearing capacity of specimens was decreased with the increasing aspect ratio.

(a) W1   (b) W2

Figure 8. *Cont.*

(c) W3   (d) W4

(e) W5

Figure 8. Failure mode of specimens: (**a**) W1; (**b**) W2; (**c**) W3; (**d**) W4; (**e**) W5.

Table 4. Characteristic load and displacement.

No.	W1	W2	W3	W4	W5	Unit
P_{cr}	136.5	389.7	76.7	309.9	234.8	kN
Δ_{cr}	2.49	13.60	4.19	0.80	0.83	mm
P_u	201.58	399.30	108.10	457.99	431.80	kN
Δ_u	8.995	14.905	20.990	6.960	2.970	mm
P_{cr}/P_u	0.677	0.975	0.709	0.676	0.543	—

Key: P_{cr} is the cracking point; Δ_{cr} is the cracking displacement; P_u is the peak load; Δ_u is the peak displacement.

The ultimate load and the cracking load of W4 and W5 exhibited about a 127%, 127%, 72%, and 114% increase compared with W1, respectively, indicating that the constraint imposed by the structural columns could significantly improve the bearing capacity of the walls. Besides that, the cracking load and ultimate load of the W5 were slightly lower than that of the W4, reflecting the bearing capacity of the ordinary concrete structural column was higher than that of the RAC structural column.

4. Test Analysis

4.1. Hysteresis Behaviour

The recorded hysteresis curves of all the specimens under cyclic loading are presented in Figure 9. Based on Figure 9a–c, the hysteresis curves of the specimens W1, W2, and W3 are considerably linear in the initial elastic stage. After the walls cracked, the area of the three hysteresis curves slightly increased, and the shapes of the hysteresis curves transformed from the original shuttle to the reversed "S" shape, showing a partial pinching effect. The area of the hysteretic curves of W1 and W3 nearly stayed the same due to the flexural effect, and the profile of the two hysteresis curves remained stable until the specimens failed. In addition, the area of the three hysteresis curves showed a marked difference.

The area of the hysteresis loop of W2 was the best, followed by that of W1, while W3 was the worst. For W1 and W2, the specimen W2 with higher compression stress had a better energy dissipation compared to W1, indicating that the increase in compressive stress can improve the energy dissipation capacity of specimens. However, for W2 and W3, specimen W3, with a higher aspect ratio, had a worse energy dissipation compared with W2, indicating that the increase in aspect ratio can reduce the energy dissipation capacity of specimens.

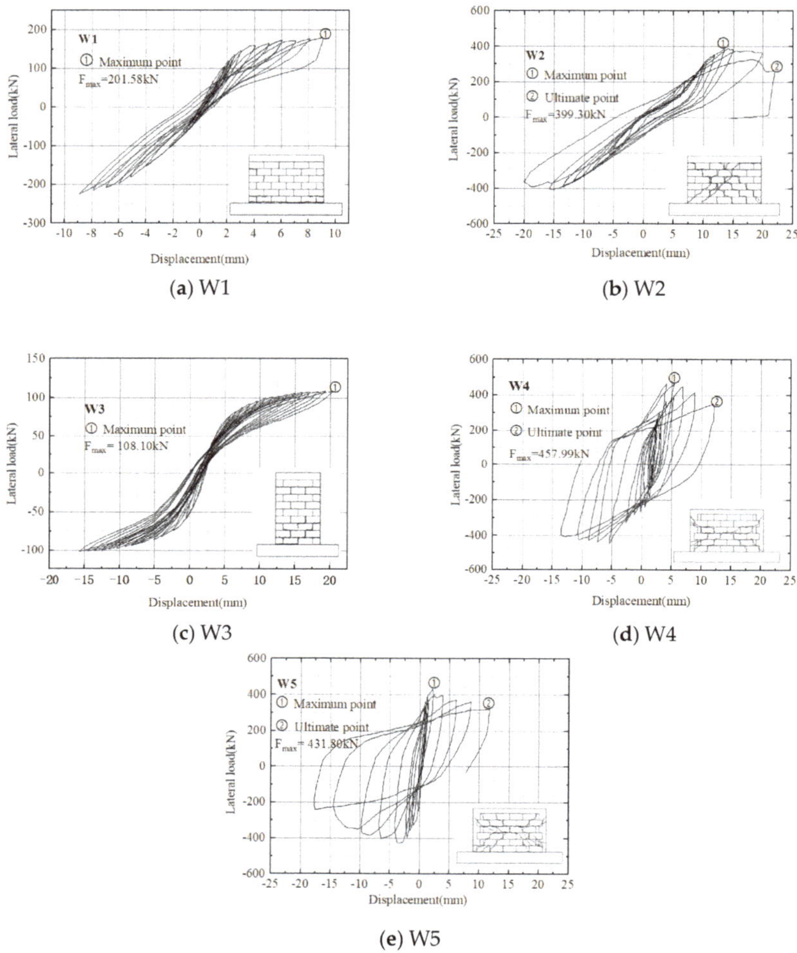

(a) W1　　(b) W2　　(c) W3　　(d) W4　　(e) W5

Figure 9. Hysteretic curves of five specimens: (**a**) W1; (**b**) W2; (**c**) W3; (**d**) W4; (**e**) W5.

As shown in Figure 9d,e, the hysteresis loops of W4 and W5 have a similar linear relationship in the initial elastic stage. During this time, the horizontal drift of the top walls and residual deformation were small after unloading. The area of the hysteresis loops of the specimens was small and nearly overlapped. After the walls cracked and entered into a plastic stage, especially after the maximum load, the hysteresis loops showed partial pinch effect, and the area of hysteresis loops was significantly increased, which indicated that the structural columns had an apparent energy dissipation capacity. Compared with W4, the maximum bearing capacity of W5 was slightly lower than that of W4, which indicated that the increasing bearing capacity of RAC structural columns was lower than that of ordinary concrete structural columns.

4.2. Skeleton Curve

The skeleton curve is the envelope curve obtained by connecting the peak points of the P–D hysteretic loop of the first cycle in each loading stage, mainly reflecting the cracking load and ductility of the wall.

As can be noticed in Figure 10a, before the specimen crack, the three skeleton curves of the specimens are a straight line in the elastic stage. After the specimens cracked, the three skeleton curves started becoming nonlinear in the plastic stage. When the lateral load closed to the maximum load, the skeleton curves gradually tilted toward the axis with increasing lateral drift. Besides that, the three skeleton curves show an apparent difference. Comparing the specimens W1 and W2, the maximum load of W2 was higher than that of W1, and the specimen W2 had a steeper degraded section after cracking, indicating that increasing the compression stress can increase the bearing capacity and decrease the ductility of specimens. The cracked load and maximum load of W3 were lower compared to W2, and the horizontal section of W3 was longer after cracking, which indicated that the increasing aspect ratio could reduce the bearing capacity and increase the ductility.

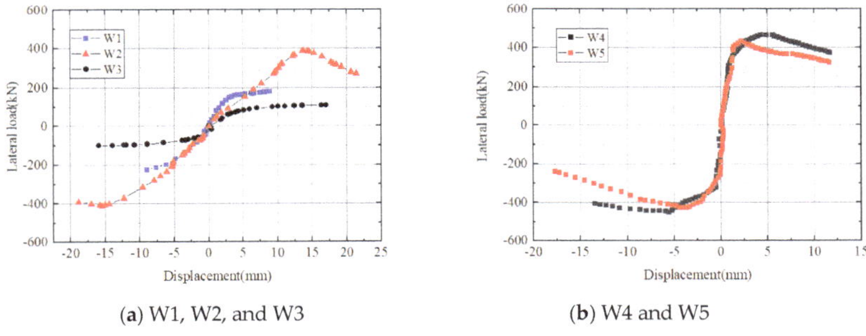

(a) W1, W2, and W3   (b) W4 and W5

Figure 10. Skeleton curves of specimens: (**a**) W1, W2, and W3; (**b**) W4 and W5.

It can be seen from Figure 10b, the bearing capacity and the deformation property of the specimens W4 and W5 are obviously higher than that of W1, W2, and W3. The stiffness degrades of the two specimens W4 and W5 was similar before cracking. After cracking, the two skeleton curves showed obvious differences in the elastic-plastic and failure phases. The maximum load of the specimen W4 was higher in comparison to the specimen W5, and the specimen W4 had a relatively smooth descending section, which reflected that the specimen W4 with ordinary concrete structural columns had the better bearing capacity and ductility than that of the specimen W5 with RAC structural columns.

4.3. Ductility

The ductility coefficient and shift angle are important parameters to assess the seismic performance of the structure, and they also are the essential characteristics to evaluate the deformation ability of the specimen.

Nowadays, the calculation methods for ductility are different, and each method has its characteristics. In this paper, the displacement ductility coefficient ($\mu = \frac{\Delta_u}{\Delta_y}$) is used for representing the ductility, where Δ_y is the yield displacement, which is obtained by using the equivalent energy method [33], Δ_u is the displacement corresponding to the lateral load of 85% of the ultimate load. The Shift angle is calculated as $\theta = \frac{\Delta_u}{H}$, where H is the distance from the top surface of the bottom beam to the lateral loading point. The ductility coefficient and shift angle of each wall are shown in Table 5.

Table 5. Ductility coefficients and shifts angle of specimens.

No.	W1	W2	W3	W4	W5	Unit
Δ_y	4.333	12.510	6.721	1.416	0.926	mm
Δ_u	8.995	14.905	20.990	6.960	2.970	mm
H	1200	1200	1800	1200	1200	mm
μ	2.075	1.191	3.123	4.915	3.207	—
θ	0.007	0.012	0.011	0.006	0.002	—

As can be seen from Table 5, the mean values of ductility coefficient μ of bare walls are between 1.191 and 3.123, and the mean values of ductility coefficient of W2 are smallest. Compared to W2, the increase in ductility was 42.6% and 162.2% for the specimens W1 and W3, indicating that increasing the compression stress could decrease the ductility of specimens, but increasing the aspect ratio could increase the ductility of specimens.

The ductility coefficients of reinforced walls were higher than 3, meeting the specification requirements of GB 50003-2011 [34] ($\mu \geq 3$), which indicated that the structural columns enhanced the deformation ability of both walls, and decreased the brittleness of specimens. According to the results of Table 5, the ductility coefficient and the lateral shift angle of W4 were higher than that of W5, which reflected that the ductility of the ordinary concrete structural columns was better than that of the RAC structural columns.

4.4. Stiffness Degradation

The stiffness degradation factor (K_i) is expressed as follows:

$$K_i = \frac{|P_i| + |-P_i|}{|\Delta_i| + |-\Delta_i|},$$

In which P_i is the maximum load and Δ_i is the corresponding displacement. The stiffness degradation of all specimens is shown in Figure 11.

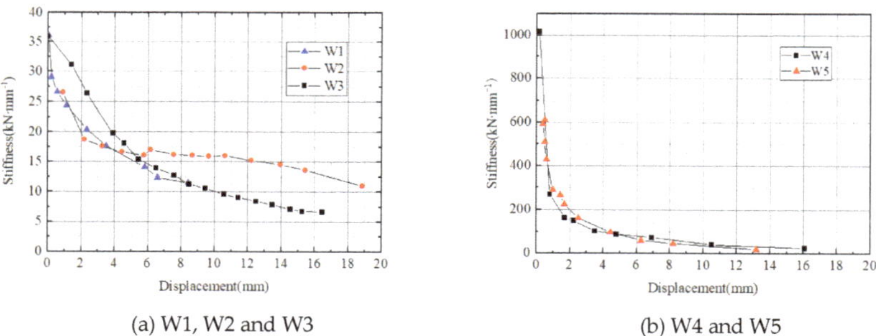

(a) W1, W2 and W3   (b) W4 and W5

Figure 11. Stiffness deterioration curves of specimens: (**a**) W1, W2, and W3; (**b**) W4 and W5.

The stiffness degradation tendency of all the specimens is plotted in Figure 11. This figure shows that the stiffness of bare walls gradually decreases with the increase of the displacement. After the displacement reached the ultimate displacement, the stiffness of bare walls degraded quickly due to the accumulation of cracks. It can be easily observed that the stiffness degradation rate of W2 was lower than that of W1 and W3 after cracking. In comparison with W1, the stiffness of W2 generally degraded, which indicated that the increase of vertical compression could reduce the degradation rate of the stiffness of specimens. In addition, when other factors were equal, a comparison of the stiffness degradation rate of W2 and W3 showed that the aspect ratio had a negative influence on the degradation rate of the stiffness of specimens.

It can be seen from Figure 11b that the stiffness degradation tendency of W4 nearly coincides with that of W5. In the initial stage, the stiffness of the specimens decreases quickly with the increase of the displacement. After the displacement reached the ultimate displacement, the stiffness degradation of reinforced specimens tended to be more smooth. The analysis stated that the effective constraint applied by structural columns could significantly alleviate stiffness degradation. In comparison with W5, the stiffness of W4 deteriorated rapidly at an early stage and deteriorated slowly at a late stage, but in general, the stiffness curves of W4 and W5 were nearly overlapping, indicating that the stiffness degradation of the specimen with RAC structural columns was close to that of ordinary concrete structural columns.

Table 6 lists the secant stiffness of characteristic points. It can be found that the stiffness degradation rate of W4 was higher than that of W5 before cracking, and after cracking, the stiffness degradation rate of W4 was slightly lower than that of W5.

Table 6. Secant stiffness of all the specimens at characteristic points.

Specimen	Secant Stiffness (kN·mm^{-1})		
	Initial Point	**Cracking Point**	**Ultimate Point**
W1	79.38	52.53	21.95
W2	60.67	44.51	14.12
W3	35.95	24.76	6.04
W4	1017.44	234.25	25.26
W5	611.59	251.24	22.53

4.5. Energy Dissipation Capacity

Energy dissipation performance refers to the energy dissipation capacity of the wall under the action of the earthquake load, which is obtained by calculating the area enclosed by the overall hysteresis loop of the first loading cycle. It is an important indicator to measure the seismic performance of the structure. The curves of energy dissipation per cycle are presented in Figure 12.

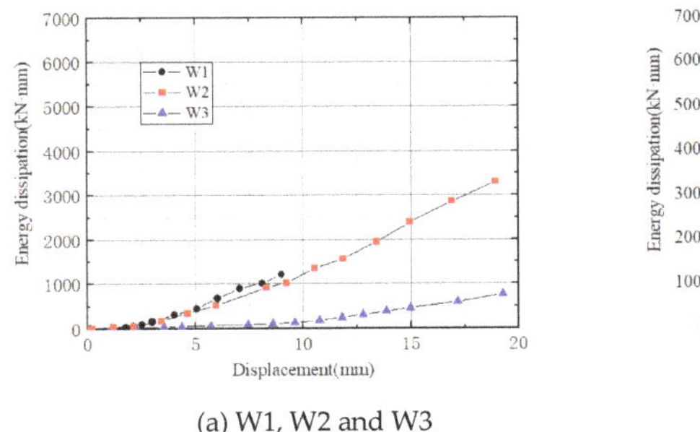

(a) W1, W2 and W3

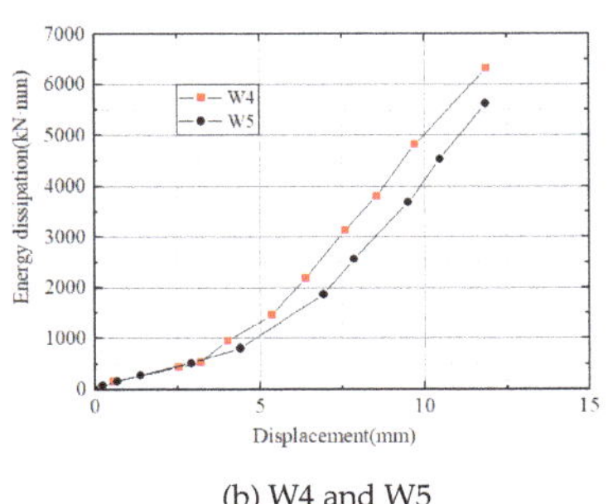

(b) W4 and W5

Figure 12. Energy dissipation per cycle for all test specimens: (**a**) W1, W2, and W3; (**b**) W4 and W5.

As shown in Figure 12a, the energy dissipation increases with the increasing lateral displacement, and the energy dissipation of W2 was the best, then the next was W1, and the worst was W3. Comparing of W1 and W2, the energy dissipation capacity of W2 was higher, indicating that, within a certain range, the vertical compression had a positive influence on the energy dissipation capacity of the specimens. The energy dissipation of W3 was lower compared with W2, indicating that increasing the aspect ratio could decrease the energy dissipation capacity of specimens.

The energy dissipation curves of the specimens W4 and W5 are shown in Figure 12b. The specimens W4 and W5 were found to have a higher energy dissipation in comparison with the bare walls, indicating that the structural columns could significantly increase the energy dissipation capacity of walls.

To further quantify the hysteretic energy dissipation performance of the specimens, the equivalent viscous damping coefficient is used to measure the energy dissipation capacity of the structure under earthquake resistance. $h_{e,u}$ and $h_{e,f}$ are the equivalent viscous damping coefficients corresponding to the maximum load and the ultimate load, respectively. The equivalent viscous damping can be calculated as follows:

$$h_e = \frac{1}{2\pi} \cdot \frac{S_{ABC}}{S_{\Delta OBD}}$$

where $S_{\Delta BOD}$ and S_{ABC} are the areas enclosed by the shaded hysteresis loop in Figure 13, respectively.

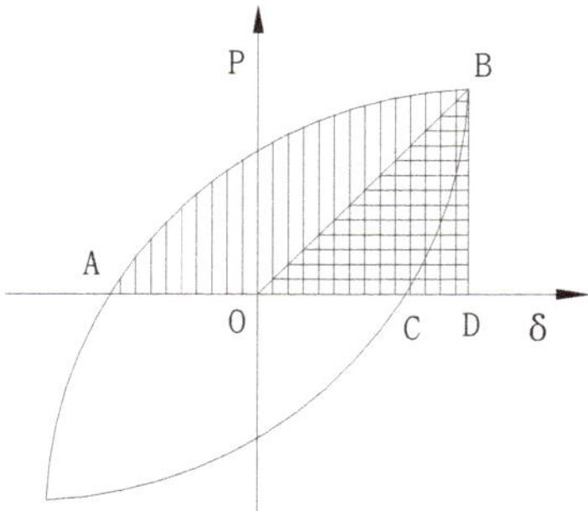

Figure 13. Calculated graphics of the equivalent viscous damping coefficient.

The value of the equivalent viscous damping coefficient and energy dissipation per cycle can be observed from Table 7 and Figure 13, respectively. In comparison with W1, the value of $h_{e,u}$ of W2 was 18.6% higher than that of W1, which indicated that the vertical compression could increase the energy dissipation performance of specimens. In terms of W2 and W3, the value of $h_{e,u}$ of W3 was 54.8% lower than that of W2, which illustrated that increasing the aspect ratio had a negative influence on the energy dissipation performance of specimens. It should be noted that W1 and W3 were almost destroyed after reaching the ultimate load, and the hysteresis curve of W2 was not closed when the specimens were destroyed. Therefore, the value of $h_{e,f}$ of W1, W2, and W3 was ignored.

Table 7. Equivalent viscous damping coefficients of all specimens.

No.	W1	W2	W3	W4	W5
$h_{e,u}$	0.087895	0.104263	0.047126	0.240740	0.210634
$h_{e,f}$	—	—	—	0.312500	0.285784

It can be easily found that the value of equivalent viscous damping ratio of W4 and W5 were more significant than that of bare walls, which illustrated that, whether each of the specimens were constrained by ordinary concrete structural columns or RAC structural columns, the equivalent viscous damping ratio and the energy dissipation of the walls were improved. Compared to W4, the energy dissipation viscous damping ratio of W5 exhibited a 12.5% and 8.9% increase at maximum load and ultimate load, which illustrated that the energy dissipation capacity of ordinary concrete structural columns was close to that of RAC structural columns.

5. Conclusions

This paper aims to develop a new concrete hollow block with RCA for the structural element. A laboratory test was carried out to investigate the seismic performance of RCHB masonry walls. Based on the experiment data and discussion, the following conclusions can be drawn:

1. In this paper, waste concrete of construction waste is re-utilized as the resource in manufacturing RCHB. This block can be used in masonry structures with seismic requirements according to the

Chinese standard. The test results indicate that this RCHB masonry structure complies with the seismic requirement through rational design. The above work provides a new solution for the recycling of construction waste;

2. RCHB masonry walls with structural columns, whether the structural columns are ordinary concrete or recycled concrete, can effectively form a constraint on the wall. The resulting constraints improve the bearing capacity and energy dissipation properties of RCHB masonry walls by more than 50%;

3. The ductility of RCHB masonry bare walls is lower than that of the RCHB masonry reinforced wall, indicating that the structural columns can significantly increase the ductility of specimens. The degradation trends of the stiffness of all bare walls are similar. Higher compressive stress can slightly increase the initial stiffness of RCHB masonry walls but accelerates the stiffness degradation after the specimens yielded. The stiffness degradation of all reinforced walls is similar;

4. With the increase of the vertical compressive stress, the bearing capacity and energy dissipation properties of the walls are improved, but the ductility of the recycled concrete block wall is decreased. With the increase of the aspect ratio, the ductility of the wall decreased, but the bearing capacity and the energy dissipation performance of specimens decreased;

5. Although the seismic performance of the RAC structural column is slightly inferior to that of the ordinary concrete structural column, the application of them on the masonry walls can significantly improve the strength and energy dissipation capacity of the RCHB masonry structure. Moreover, compared with the ordinary concrete structural column, the RAC structural column is more economical and environmentally friendly.

Author Contributions: Writing—review and editing, C.L. and X.N.; Resources, F.Z., Z.Q. and G.B.

Funding: This research was funded by [the Natural Science Foundation of China] grant number [51878546], [the Shaanxi Province Innovative Talent Promotion Plan Project] grant number [2018KJXX-056], [the Shaanxi Provincial Science and Technology Innovation Planning Project] grant number [2016KTZDSF04-02-01], the Shaanxi Provincial Key Research and Development Program] grant number [2018ZDCXL-SF-03-03-02], [the Shaanxi Provincial Key Technology and Science Innovation Team Program] grant number [2016-KCT31], [the Shaanxi Science and Technology Innovation Base] grant number [2017KTPT-19].

Acknowledgments: The support from the following fund projects is highly acknowledged: (1) the National Natural Science Foundation of China (51878546); (2) the Shaanxi Provincial Innovative Talents Promotion Program (2018KJXX-056); (3) the Shaanxi Provincial Science and Technology Innovation Planning Project (2016KTZDSF04-02-01); (4) the Shaanxi Provincial Key Research and Development Project (2018ZDCXL-SF-03-03-02); (5) the Shaanxi Provincial Key Technology and Science Innovation Team Program (2016-KCT31); and (6) the Shaanxi Science and technology innovation base "technology innovation Service Platform of solid waste resources energy-saving wall materials science" (2017KTPT-19).

Conflicts of Interest: The authors declare no conflict of interest.

References

1. Ibrahim, M.I.M. Estimating the sustainability returns of recycling construction waste from building projects. *Sustain. Cities Soc.* **2016**, *23*, 78–93. [CrossRef]

2. Bravo, M.; De Brito, J.; Pontes, J.; Evangelista, L. Mechanical performance of concrete made with aggregates from construction and demolition waste recycling plants. *J. Clean. Prod.* **2015**, *99*, 59–74. [CrossRef]

3. Yang, H.; Xia, J.Q.; Thompson, J.R.; Flower, R.J. Urban construction and demolition waste and landfill failure in Shenzhen, China. *Waste Manag.* **2017**, *63*, 393–396. [CrossRef] [PubMed]

4. Xu, J.P.; Shi, Y.; Zhao, S.W. Reverse Logistics Network-Based Multiperiod Optimization for Construction and Demolition Waste Disposal. *J. Constr. Eng. Manag.* **2018**, *2*, 145. [CrossRef]

5. He, S.Y.; Lai, J.X.; Wang, L.X.; Wang, K. A literature review on properties and applications of grouts for shield tunnel. *Constr. Build. Mater.* **2020**, *231*, 468–482.

6. Feng, S.J.; Shi, Z.M.; Shen, Y.; Li, L.C. Elimination of loess collapsibility with application to construction and demolition waste during dynamic compaction. *Environ. Earth. Sci.* **2015**, *73*, 5317–5332. [CrossRef]

7. Song, Q.B.; Li, J.H.; Zeng, X.L. Minimizing the increasing solid waste through zero waste strategy. *J. Clean. Prod.* **2015**, *104*, 199–210. [CrossRef]

8. Huang, B.J.; Wang, X.Y.; Kua, H.W.; Geng, Y.; Bleischwitz, R.; Ren, J.Y. Construction and demolition waste management in China through the 3R principle. *Resour. Conser. Recycl.* **2018**, *129*, 36–44. [CrossRef]

9. Li, P.L.; Lu, Y.Q.; Lai, J.X.; Liu, H.Q. A comparative study of protective schemes for shield tunneling adjacent to pile groups. *Adv. Civ. Eng..* (In press).

10. Lv, Z.Y.; Liu, C.; Zhu, C.; Bai, G.L.; Qi, H. Experimental Study on a Prediction Model of the Shrinkage and Creep of Recycled Aggregate Concrete. *Appl. Sci.* **2019**, *9*, 4322. [CrossRef]

11. Wu, H.Y.; Duan, H.B.; Zheng, L.; Wang, J.Y.; Niu, Y.N.; Zhang, G.M. Demolition waste generation and recycling potentials in a rapidly developing flagship megacity of South China: Prospective scenarios and implications. *Constr. Build. Mater.* **2016**, *113*, 1007–1016. [CrossRef]

12. Xiao, J.Z. *Recycled Concrete*; China Architecture & Building Press: Beijing, China, 2008.

13. Sonagnon, M.; Ahmed, Z.; Bendimerad, E. How do recycled concrete aggregates modify the shrinkage and self-healing properties. *Cem. Concr. Compos.* **2018**, *86*, 72–86. [CrossRef]

14. Etxeberria, M.; Vázquez, E.; Marí, A.R. Microstructure analysis of hardened recycled aggregate concrete. *Mag. Concr. Res.* **2006**, *586*, 83–90. [CrossRef]

15. Duan, Z.H.; Poon, C.S. Properties of recycled aggregate concrete made with recycled aggregates with different amounts of old adhered mortars. *Mater. Des.* **2014**, *58*, 19–29. [CrossRef]

16. De Juan, M.S.; Gutiérrez, P.A. Study on the influence of attached mortar content on the properties of recycled concrete aggregate. *Constr. Build. Mater.* **2009**, *23*, 872–877. [CrossRef]

17. Corinaldesi, V.; Letelier, V.; Moriconi, G. Behaviour of beam–column joints made of recycled-aggregate concrete under cyclic loading. *Constr. Build. Mater.* **2011**, *25*, 1877–1882. [CrossRef]

18. Jelić, I.; Šljivić-Ivanović, M.; Dimovic, S.; Antonijević, D.; Jović, M.; Mirković, M.; Smičiklas, I. The applicability of construction and demolition waste components for radionuclide sorption. *J. Clean. Prod.* **2018**, *171*, 322–332. [CrossRef]

19. Liu, C.; Liu, H.W.; Zhu, C.; Bai, G.L. On the mechanism of internal temperature and humidity response of recycled aggregate concrete based on the recycled aggregate porous interface. *Cem. Concr. Compos.* **2019**, *103*, 22–35. [CrossRef]

20. Nie, X.X.; Feng, S.S.; Zhang, S.D.; Gan, Q. Simulation study on the dynamic ventilation control of single head roadway in high-altitude mine based on thermal comfort. *Adv. Civ. Eng.* **2019**, *2019*, 12. [CrossRef]

21. Liu, C.; Fan, Z.Y.; Chen, X.N.; Zhu, C.; Wang, H.Y.; Bai, G.L. Experimental Study on Bond Behavior between Section Steel and RAC under Full Replacement Ratio. *KSCE J. Civ. Eng.* **2018**, *23*, 1159–1170. [CrossRef]

22. Liu, C.; Lv, Z.Y.; Bai, G.L.; Yin, Y.G. Experiment study on bond slip behavior between section steel and RAC in SRRC structures. *Constr. Build. Mater.* **2018**, *175*, 104–114. [CrossRef]

23. Liu, C.; Lv, Z.Y.; Bai, G.L.; Zhang, Y. Study on Calculation Method of Long-Term Deformation of RAC Beam based on Creep Adjustment Coefficient. *KSCE J. Civ. Eng.* **2019**, *23*. [CrossRef]

24. Soutsos, M.N. Concrete building blocks made with recycled demolition aggregate. *Constr. Build. Mater.* **2011**, *25*, 726–735. [CrossRef]

25. Bai, G.L.; Zhang, F.J.; An, Y.F.; Xiao, H.; Quan, Z.G. Experimental study on compression study of recycled concrete hollow block masonry. *J. Xi'an Univ. Arch. Technol.* **2011**, *43*, 7–17. (In Chinese)

26. Poon, C.S.; Kou, S.C.; Lam, L. Use of recycled aggregates in molded concrete bricks and blocks. *Constr. Build. Mater.* **2002**, *16*, 281–289. [CrossRef]

27. Poon, C.S.; Kou, S.C.; Wan, H.W.; Etxeberria, M. Properties of concrete blocks prepared with low grade recycled aggregates. *Waste Manag.* **2009**, *23*, 2877–2886. [CrossRef]

28. Guo, Z.G.; Tu, A.; Chen, C.; Lehman, D.E. Mechanical properties, durability, and life-cycle assessment of concrete building blocks incorporating recycled concrete aggregates. *J. Clean. Prod.* **2018**, *199*, 136–149. [CrossRef]

29. Sabai, M.M.; Cox, M.G.D.M.; Mato, R.R.; Egmond, E.L.C.; Lichtenberg, J.J.N. Concrete block production from construction and demolition waste in Tanzania. *Resour. Conser. Recycl.* **2013**, *72*, 9–19. [CrossRef]

30. Corinaldesi, V. Mechanical behavior of masonry assemblages manufactured with recycled-aggregate mortars. *Cem. Concr. Compos.* **2009**, *31*, 505–510. [CrossRef]

31. Ministry of Housing and Urban-Rural Development of the People's Republic of China. *Test Methods for the Concrete Block and Brick*; GB/T 41112013; Chinese Building Press: Beijing, China, 2013.

32. Ministry of Housing and Urban-Rural Development of the People's Republic of China. *Normal Concrete Small Block*; GB/T 8239-2014; Chinese Building Press: Beijing, China, 2014.
33. Park, R. Ductility evaluation from laboratory and analytical testing. In Proceedings of the 9th World Conference on Earthquake Engineering, Tokyo–Kyoto, Japan, 2–8 August 1988; pp. 605–616.
34. Ministry of Housing and Urban-Rural Development of the People's Republic of China. *Code for Design of Masonry Structures*; GB 50003-2011; Chinese Building Press: Beijing, China, 2011. (In Chinese)

© 2019 by the authors. Licensee MDPI, Basel, Switzerland. This article is an open access article distributed under the terms and conditions of the Creative Commons Attribution (CC BY) license (http://creativecommons.org/licenses/by/4.0/).

Article

Experimental Study on Seismic Behavior of Steel Frames with Infilled Recycled Aggregate Concrete Shear Walls

Lijian Sun [1,2]**, Hongchao Guo** [1,2,*] **and Yunhe Liu** [1,2]

[1] State Key Laboratory of Eco-hydraulics in Northwest Arid Region, Xi'an University of Technology No.5 Jinhua Road, Xi'an 710048, China; sunlijian0414@163.com (L.S.); liuyunhe1968@163.com (Y.L.)

[2] School of Civil Engineering and Architecture, Xi'an University of Technology No.5 Jinhua Road, Xi'an 710048, China

* Correspondence: ghc_1209@163.com; Tel.: +86-132-2700-9454

Received: 16 October 2019; Accepted: 4 November 2019; Published: 5 November 2019

Abstract: Experiments were performed on four specimens of steel frames with infilled recycled aggregate concrete shear walls (SFIRACSWs), one specimen of infilled ordinary concrete wall, and one pure-steel frame were conducted under horizontal low cyclic loading. The influence of the composite forms of steel frames and RACSWs (namely, infilled cast-in-place and infilled prefabricated) on the failure modes, transfer mechanisms of lateral force, bearing capacity, and ductility of SFIRACSWs is discussed, and the concrete type and connecting stiffness of beam–column joints (BCJs) are also considered. Test results showed that infilled RACSWs can increase the bearing capacity and lateral stiffness of SFIRACSWs. The connecting stiffness of BCJs slightly influences the seismic behavior of SFIRACSWs. In the infilled cast-in-place RACSWs, the wall cracks mainly extended along the diagonal direction. The bearing capacity was 2.4 times higher than in the pure steel frame, the initial stiffness was 4.3 times higher, and the displacement ductility factors were 2.44–2.69 times higher. In the infilled prefabricated RACSWs, the wall cracks mainly extended along the connection between the embedded T-shape connectors and walls before finally connecting along the horizontal direction. Moreover, shear failure occurred in the specimens. The bearing capacity was 1.44 times higher than that of the pure steel frame, the initial stiffness was 2.8 times higher, and the displacement ductility factors were 3.32–3.40 times higher. The degradation coefficients of the bearing capacity were more than 0.85, indicating that the specimens demonstrated a high safety reserve.

Keywords: steel frame; infilled shear walls; recycled aggregate concrete; semi-rigid connection; seismic behavior

---

Highlights

- The practicability of recycled aggregate concrete shear walls (RACSWs) as lateral resistance components of steel structures is investigated.
- An experiment on steel frames with infilled cast-in-place and prefabricated RACSWs (SFIRACSWs) was conducted.
- The effects of concrete type, composite forms of steel frames and RACSWs, and connecting stiffness of beam–column joints are considered.
- The failure modes and transfer mechanisms of lateral force of SFIRACSWs are clarified.
- The main seismic performance indexes of SFIRACSWs are compared with those of pure steel frames.

1. Introduction

Recycled aggregate concrete (RAC) can fundamentally solve concrete waste problems not only by reducing environmental pollution from waste concrete, but also preserving natural aggregates and reducing the consumption of natural resources and energy. RAC is one of the main approaches to developing a circular economy and promoting environmentally friendly buildings. The suitability of waste concrete as recycled aggregate for construction projects has been investigated. Puthussery et al. [1] reported that recycled aggregate can be used as a building material for road construction, mass concrete engineering, and lightly reinforced sections, thereby providing ideas for recycling concrete waste.

With regard to the mechanical properties of RAC, the compressive and tensile strengths of recycled coarse aggregate concrete with different sources and strength grades were studied. Tabsh et al. [2], Koenders et al. [3], and Silva et al. [4] found that the strength reduction of RAC is clearer with low-strength coarse aggregate than with high-strength aggregate, and the compressive and tensile strengths of RAC made of 50 MPa coarse aggregate are equal to those of natural aggregate concrete. Bairagi et al. [5] and Oikonomou [6] proposed stress–strain curves of RAC with different aggregate replacement rates in which the constitutive relation of RAC with different aggregate replacement rates was similar, and only the decline stage was different. Ying et al. [7] and Wang et al. [8] studied the diversity of chloride ion diffusion in RAC and analyzed the influence of carbonation modification on the interface properties of RAC. Carbonation can improve the interface properties of RAC, especially when the water–cement ratios of new and old cement mortars are high; thus, the improvement effect is noticeable.

With respect to the components of RAC, Arezoumandi et al. [9] and Choi et al. [10] tested the shear strength of RAC beams under short- and long-term loads and discussed the applicability of the design code, modifying compression field theory to the shear strength of the RAC beams. The axial compressive performance of RAC columns was also investigated. Choi et al. [11] and Xiao et al. [12] found that the maximum axial compressive bearing capacity of RAC columns decreases slightly with an increase in the replacement rate of recycled coarse aggregate, and RAC columns can be used for the load-bearing member of the structure. Wu et al. [13,14] proposed a concept for recycling mixed components based on the recycling technology of large-scale waste concrete blocks and systematically studied thin-walled steel tubular columns, U-shape steel beams, and thin steel-plate walls filled with RAC. By testing RAC-filled square steel tube (RACFST) columns and RACFST columns strengthened by carbon-fiber-reinforced polymer, Chen et al. [15] and Dong et al. [16] found that the RACFST columns exhibited good seismic performance under low axial compression; the aggregate replacement rate demonstrated a minimal influence on the RACFST columns. Fathifazl et al. [17] and Ma et al. [18] studied steel-reinforced RAC (SRRAC) beams and columns and investigated the effect of the replacement rates of recycled coarse aggregate, axial compression ratios, and stirrup ratios on the seismic performance of the SRRAC columns.

For the RAC structures, the seismic performance of two connections was compared and analyzed by testing the RAC beam–column joints (BCJs) [19,20]. Xiao et al. [21] and Wang et al. [22] performed the shaking table test on RAC frames and reported the dynamic response of RAC frames. Tests on RAC shear walls (RACSWs) have been conducted. Peng et al. [23] and Ma et al. [24] found that the existing formulas cannot predict the peak load and failure modes of squat RACSWs and proposed a mixed flexural and diagonal compression mechanism. The dynamic responses and failure modes of ordinary concrete and RAC specimens were compared and discussed by using the shaking table test on RAC frame-shear walls [25].

In general, recycled coarse aggregate is different from natural gallet and pebble aggregate, and its porosity is high, thereby resulting in high water absorption, large dry shrinkage and creep, and poor bond performance of the RAC. Considerable research on the mechanical properties, components, and structures of RAC have demonstrated that RAC shows similar mechanical properties to ordinary concrete and can be widely used in construction after rational design. The main load-bearing components of RAC structures include recycled coarse aggregates, which were popularized during the construction of towns after the Wenchuan earthquake in 2008.

Steel structure residences have the advantages of strong seismic resistance, high industrialization, recyclability, and reduced resource consumption and construction waste discharge. They are also among the residential structure systems preferred by developed countries at present. In addition to using steel to build steel structure residences, the material development and application technology of enclosure systems have also been assessed. This scenario is a technical problem for new building energy-saving materials and systems, and a social and economic development problem for factory construction of housing in terms of changing construction modes. Tong et al. [26] and Sun et al. [27] tested semi-rigid steel frame-filled concrete shear walls and determined that the energy dissipation of this composite structure mainly depends on the aggregate friction and occlusion between the cracks of filler walls and the yield of shear studs. The structure has multiple transmission paths of horizontal loads and a high safety reserve. Kurata et al. [28] and Guo et al. [29,30] conducted research on a steel-plate shear wall with a semi-rigid steel frame and found that the structure exhibits the advantages of semi-rigid joints with good rotation capability as well as energy absorption through the yield deformation of the steel plate. The hysteretic performance is stable and characterized by the simplified construction and efficient utilization of materials. Wu et al. [31] conducted an experimental study on steel frames with replaceable reinforced concrete walls. Other tests, such as tests on steel frames with fabricated autoclaved lightweight concrete panels, composite lightweight walls, and light-gauge steel stud walls, were conducted under horizontal low cyclic loading [32–34].

Prefabricated construction systems have the advantages of fast construction, stable quality, and energy saving, along with being an environmentally friendly and sustainable development technology. In this study, two aspects of improvements were considered on the basis of the structure of steel frames filled with concrete shear walls as proposed by the previous scholars. On the one hand, the RAC concrete was used to replace the ordinary concrete, and concrete shear walls made of recycled coarse aggregate were introduced into the steel frame structure. The walls bore the horizontal loads as the main lateral resisting components of the structure when they functioned as the enclosure, thereby providing a new idea for popularizing and applying RAC in steel structure residences. On the other hand, the prefabricated connection between steel frames and concrete shear walls was also considered to facilitate the rapid construction of the project and the timely replacement of the damaged walls. Therefore, the structure of steel frames with infilled RACSWs (SFIRACSWs) is proposed in this paper, and tests on SFIRACSWs were conducted under horizontal low cyclic loading. The effect of two composite forms of steel frames and RACSWs (namely, infilled cast-in-place and infilled prefabricated) on the mechanical behavior of SFIRACSWs is discussed. The failure modes and cooperative mechanism of steel frames and RACSWs is clarified, and the main seismic performance indexes of SFIRACSWs are evaluated comprehensively, providing a theoretical basis for popularizing and applying SFIRACSWs in practical engineering.

2. Experimental Program

2.1. Specimen Design

To analyze the influence of concrete type, the composite forms of steel frames and RACSWs, and the connecting stiffness of beam–column joints (BCJs) on the hysteretic behavior of SFIRACSWs, this study designed six specimens of one-story and one-bay, at a 1:3 scale, and divided them into three groups. The first and second groups were designed to analyze the effects of the composite forms of steel frames and RACSWs, and the third group was a pure steel frame that was used as a reference specimen.

The first group comprised specimens of infilled cast-in-place RACSWs, which were labeled SPE1, SPE2, and SPE3. Only the wall of SPE1 was made of ordinary concrete, to illustrate the effect of concrete type. The walls of the first group were poured at the construction site and connected with steel frames through shear stubs. The second group consisted of specimens of infilled prefabricated RACSWs, which were labeled SPE4 and SPE5. The walls were prefabricated in the factory and assembled rapidly with steel frames through T-shape connectors embedded in the walls. Specimen SPE6 belonged to the third group.

The span and height of the specimens were 1050 and 1200 mm, respectively; the beam section was HN $150 \times 100 \times 5 \times 8$, and the column section was HW $150 \times 150 \times 7 \times 10$. Two forms of BCJs, namely, welded–bolted rigid joint and flush end-plate semi-rigid joint, were adopted, as illustrated in Figure 1a,b, to investigate the influence of the connecting stiffness of BCJs on the seismic performance of the SFIRACSWs. Specimens SPE1, SPE2, SPE4, and SPE6 adopted rigid joints, while the other specimens used semi-rigid joints. The main parameters of the specimens are presented in Table 1, and the dimensions and details of the specimens are illustrated in Figure 1.

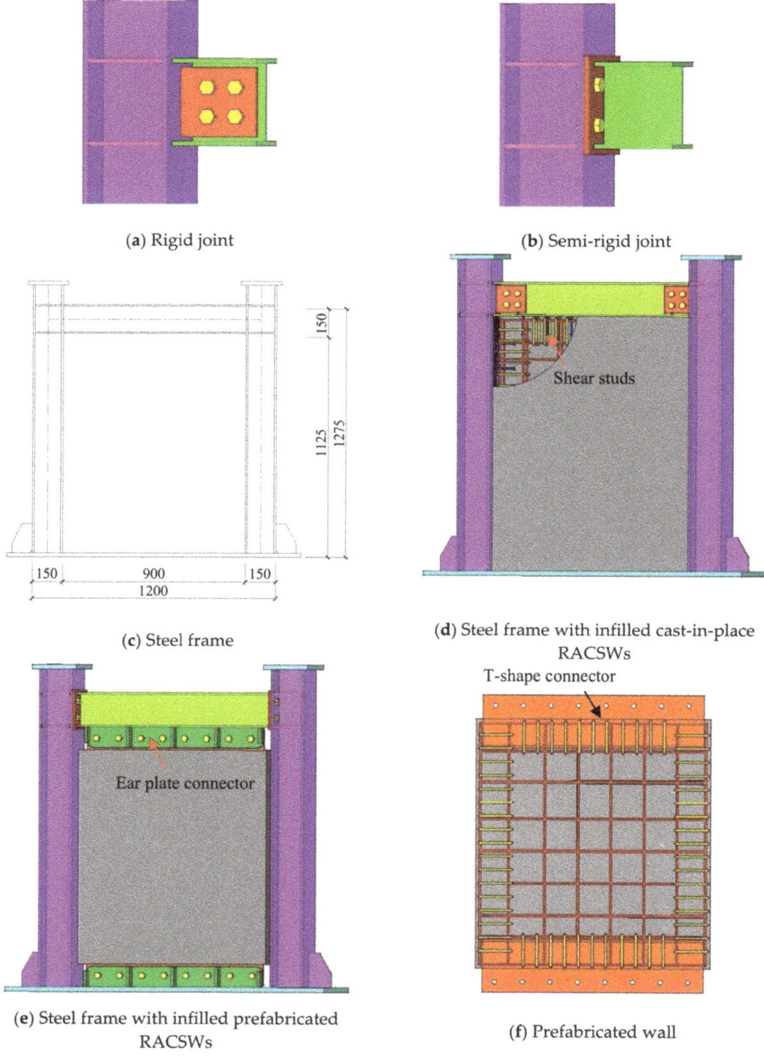

Figure 1. Dimensions and details of specimens.

Table 1. Main parameters of specimens.

Group	Specimen	Beam–Column Joints	Concrete Type	Wall Type	Connection of Walls and Steel Frames	Reinforcement of Walls (mm)
One	SPE1	Welded-bolted	ordinary concrete	cast-in-place	Infilled shear studs	Double layer, double way Φ6@120
	SPE2	Welded-bolted	RAC			
	SPE3	Flush end-plate	RAC			
Two	SPE4	Welded-bolted	RAC	prefabricated	Ear plates, T-shape connectors and bolts	Double layer, double way Φ6@120
	SPE5	Flush end-plate	RAC			
Three	SPE6	Welded-bolted	-	-	-	-

The diameter of the recycled coarse aggregate was 10–30 mm, and the replacement rate of the coarse aggregate was 100%. The cast-in-place RACSWs with a thickness of 90 mm were connected with steel frames by M16 shear studs, which were welded on the steel frames at 110-mm intervals, as depicted in Figure 1d. The 925 × 860 × 90 mm prefabricated RACSWs were connected with steel frames by ear plates, T-shape connectors, and high-strength bolts. The spacing of the bolts was 100 mm, and the rapid assembly of the walls and steel frames could be achieved as illustrated in Figure 1e. The ear plate connectors were welded on the flange of steel beams and equipped with stiffeners at 200-mm intervals. The T-shape connectors were embedded in the concrete wall and welded with the steel bars of the hidden beams (Figure 1f).

Hidden beams and columns were set around the walls. Four Φ8 steel bars were in the hidden beams and columns, and the diameter of the stirrup spacing of 50 mm was 6 mm. A double-layer steel mesh was arranged on the wall, and the diameter of the horizontal and vertical steel bars with a spacing of 120 mm was 6 mm. The reinforcement details of the infilled walls are presented in Figure 2.

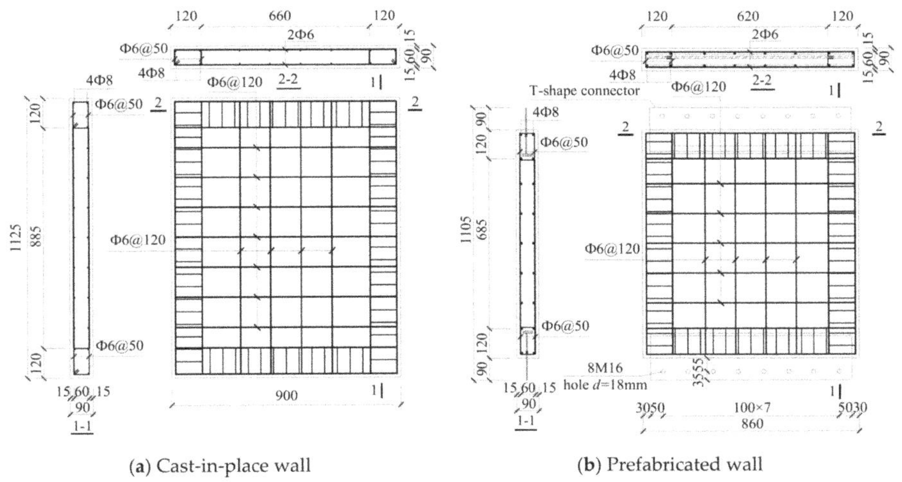

(**a**) Cast-in-place wall   (**b**) Prefabricated wall

Figure 2. Reinforcement details of infilled walls.

2.2. Material Properties

The tensile strength tests of steel were conducted according to the code *Metallic Materials—Tensile Testing—Part I: Method of Test at Room Temperature*. All steel was Q235B with a yield strength of 235 MPa. The grade of all steel bars was HPB300 with a yield strength of 300 MPa. The mechanical properties of the tested steel are listed in Table 2. The recycled coarse aggregate was from waste concrete specimens that had been placed in the laboratory for many years and were broken into concrete blocks. The design strength grade of RAC was C30 and the cubic compressive strength was 30 MPa. Test cubes with a size of 100 × 100 × 100 mm were created while pouring walls and cured under the same condition as the walls. The cubic compressive strength of the RAC was measured as 32.8 MPa according to the *Standard for Test Method of Mechanical Properties of Ordinary Concrete.*

Table 2. Mechanical properties of steel.

Interception Position	Thickness (mm)	Yield Stress (N/mm^2)	Ultimate Stress (N/mm^2)	Young's Modulus (N/mm^2)	Elongation at Fracture%
Beam flange	8	270.20	402.30	2.09×10^5	31.95
Beam web	5	302.60	413.10	2.64×10^5	35.15
Column flange	10	268.30	447.05	2.34×10^5	34.40
Column web	7	283.75	452.00	2.52×10^5	34.00
Column stiffener	8	281.55	403.95	1.80×10^5	32.85
Φ6 steel bar	-	217.30	345.50	2.50×10^5	32.70
Φ8 steel bar	-	348.34	482.37	2.62×10^5	37.65

2.3. Test Setup and Loading Procedure

The test loading device is illustrated in Figure 3. The specimen was anchored on the ground beam with M30 bolts, and both ends of the ground beam were fixed in the laboratory by pressure beams. The horizontal load was applied by a 1000-kN MTS actuator, and the vertical load was applied by a 1000-kN hydraulic jack. A lateral brace was provided at the end of the MTS actuator, and the outside displacement of the specimen was limited by two groups of pulleys to ensure that the specimen and actuator moved in the horizontal direction. The displacement and strain gauges were arranged in the key parts to study the seismic behavior of the SFIRACSWs, and data were collected by a TDS-630 acquisition instrument. Concrete cracking, buckling of beams and columns, and specimen failure were constantly observed during the test.

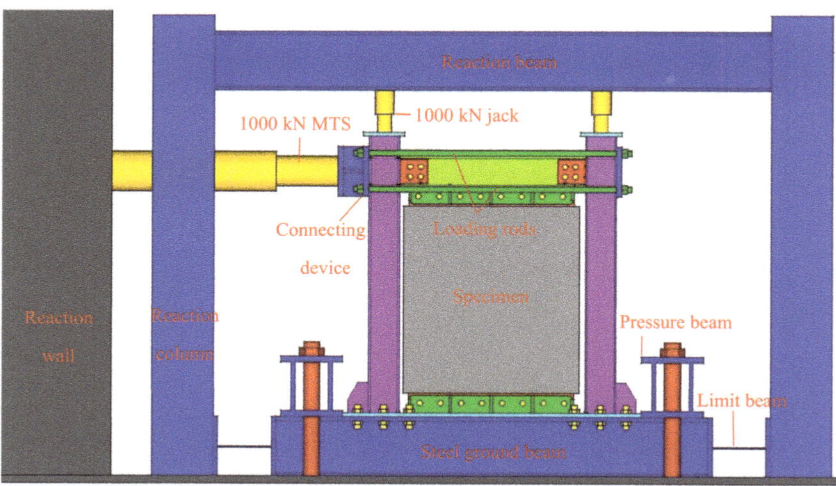

Figure 3. Test setup.

In the vertical direction, 250 kN loads were applied to the steel columns, which was calculated by an axial compression ratio of 0.3. The axial compression ratio is the ratio of the design value of the axial load to the product of the total section area and design value of the axial compressive strength of concrete. The horizontal load was applied by the joint control method of force and displacement.

Before the yield of the specimen, the load was controlled by force and cycled once with an increase of 20 kN each time. The load-displacement curves showed a noticeable turning point as a yield sign. After the yield of the specimen, the load was controlled by displacement and cycled thrice with an increase of 0.5 δ_y each time, which was the estimated yield displacement. Loading stopped when the horizontal load was down to 85% of the peak load.

3. Behavior of Test Specimens

3.1. General Behavior

3.1.1. Cast-in-Place RACSWs

(1) Specimen SPE1

Specimen SPE1 was in the elastic stage with no observable behavior when the horizontal load was less than 200 kN. Slight oblique cracks began to appear in the middle part of the western side of the wall when the load was 220 kN. Oblique cracks appeared on the lower part of the eastern side of the wall when the load was 260 kN. Then, the specimen yielded locally, and the load was applied by controlling the displacement.

In the displacement control stage, the wall cracks continued to expand, and oblique cracks formed along the 45° direction in the middle of the wall when the displacement was 1.5 δ_y (Figure 4a). The cracks continued to expand and extend at the control stage of displacement 2.0–4.0 δ_y, and principal cracks with widths of 2–3 mm gradually formed on both sides of the wall (Figure 4b). The concrete along both sides of the principal cracks began to be crushed and fall off, and the width of the cracks reached 5–8 mm (Figure 4c). Local concrete fell off on both sides of the principal diagonal cracks, and the flange at the end of the beam bulged clearly when the displacement was 5.5 δ_y (Figure 4d). A large area of concrete fell off in the middle of the wall, numerous steel bars were exposed, and the column base buckled outside when the displacement was 6.0 δ_y. At the control stage of displacement 6.5 δ_y, the wall was badly damaged and had holes in the middle. Finally, the horizontal load was reduced by more than 15%, and the specimen lost its carrying capacity.

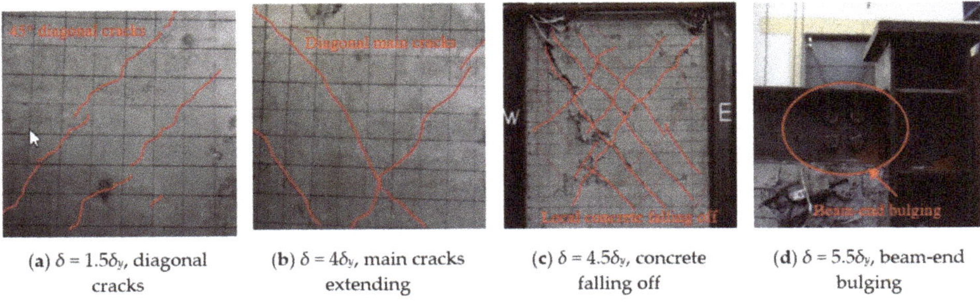

(**a**) $\delta = 1.5\delta_y$, diagonal cracks (**b**) $\delta = 4\delta_y$, main cracks extending (**c**) $\delta = 4.5\delta_y$, concrete falling off (**d**) $\delta = 5.5\delta_y$, beam-end bulging

Figure 4. Local failure of SPE1.

(2) Specimen SPE2

Specimen SPE2 was in the elastic stage with no observable behavior when the horizontal load was less than 140 kN. A slight crack appeared on the lower part of the wall when the load was 160 kN. The wall cracks continued to expand when the load was 280 kN. Then, the specimen yielded locally, and the load was applied by controlling the displacement.

In the displacement control stage, several oblique cracks were observed along the 45° direction of the wall, with a length of approximately 100 cm when the displacement was 1.5 δ_y (Figure 5a). The wall cracks continued to expand and extend to the edge of the wall when the displacement was 2.0 δ_y, thereby forming three principal cracks along the diagonal direction. The local concrete expanded and fell off at the intersection of the principal diagonal cracks when the displacement was 3.5 δ_y, and the width of the cracks reached 4–5 mm (Figure 5b). A large area of concrete on top of the wall fell off and extended along the principal cracks when the displacement was 4.5 δ_y, thereby forming two dropping areas with a length of approximately 50 cm. The local steel bars were exposed. Numerous steel bars were exposed when the displacement was 5.5 δ_y. Then, the local wall showed holes. The flange at the end of the beam bulged upward (Figure 5c) and the column base buckled (Figure 5d). Finally, the horizontal load was reduced by more than 15%, and the specimen lost its carrying capacity.

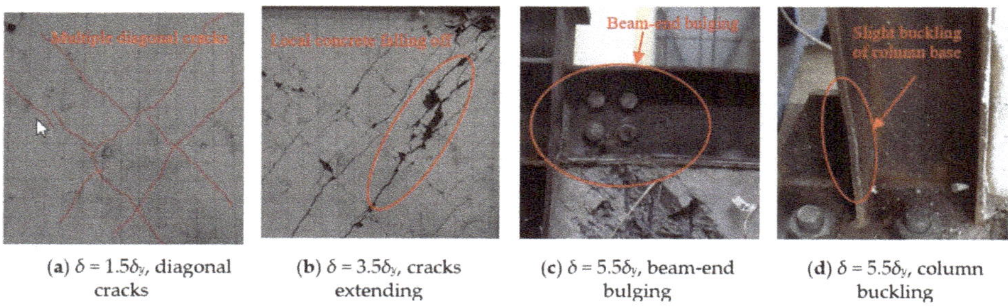

(**a**) δ = 1.5δ_y, diagonal cracks   (**b**) δ = 3.5δ_y, cracks extending   (**c**) δ = 5.5δ_y, beam-end bulging   (**d**) δ = 5.5δ_y, column buckling

Figure 5. Local failure of SPE2.

(3) Specimen SPE3

Specimen SPE3 was in the elastic stage with no observable behavior when the horizontal load was less than 180 kN. Numerous fine cracks appeared on the right side of the wall when the load was 200 kN. The cracks constantly expanded when the load was 280 kN. Then, the specimen yielded locally, and the load was applied by controlling the displacement.

The cracks at the control stage of displacement 1.0–3.0 δ_y continued to expand and extend to the edge of the wall, thereby forming through cracks with a width of 2–3 mm along the diagonal direction (Figure 6a). The concrete partly fell off on both sides of the principal diagonal cracks when the displacement was 4.0 δ_y, and the width of the cracks reached 4–5 mm (Figure 6b). Considerable concrete fell off at the intersection of the cracks along the diagonal when the displacement was 4.5 δ_y. The local steel bars were exposed. Then, the flange at the end of the beam bulged upward. A large area of concrete fell off when the displacement was 5.5 δ_y. Then, many holes appeared on the wall. Numerous steel bars were exposed and the wall was seriously damaged. Furthermore, the end-plate warped (Figure 6c). The middle part of the column bulged outward and the column base buckled (Figure 6d). Finally, the horizontal load decreased by more than 15%, and the specimen lost its carrying capacity.

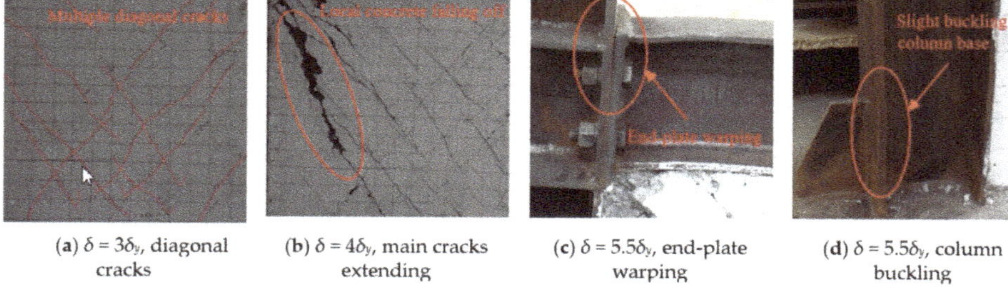

(**a**) δ = 3δ_y, diagonal cracks   (**b**) δ = 4δ_y, main cracks extending   (**c**) δ = 5.5δ_y, end-plate warping   (**d**) δ = 5.5δ_y, column buckling

Figure 6. Local failure of SPE3.

3.1.2. Prefabricated RACSWs

(1) Specimen SPE4

Specimen SPE4 was in the elastic stage with no observable behavior when the horizontal load was less than 120 kN. An initial crack appeared at the lower right corner of the wall when the load was 140 kN. Multiple fine oblique cracks appeared in the wall center and extended when the load was 220 kN. The specimen yielded locally, then the load was applied by controlling the displacement.

In the displacement control stage, multiple short cracks with a width of approximately 1 mm appeared along the diagonal direction of the wall when the displacement was 1.5 δ_y. The concrete at the lower right corner of the wall began to fall off, and the number of horizontal cracks gradually increased (Figure 7a). Cracks in the horizontal direction formed at the bottom of the wall when the displacement was 2.5 δ_y (Figure 7b). The upper ear plate had a relative slip of approximately 10 mm with the T-shape connector (Figure 7c) accompanied by a friction sound among steel plates. Bending deformation occurred on the wall when the displacement was 4.5 δ_y. Then, the corner concrete fell off, exposing the steel bars. The top flange of the steel beam exhibited a brittle fracture when the displacement was 5.0 δ_y (Figure 7d), and the column base buckled locally. Finally, the horizontal load was reduced by more than 15%, and the specimen lost its carrying capacity.

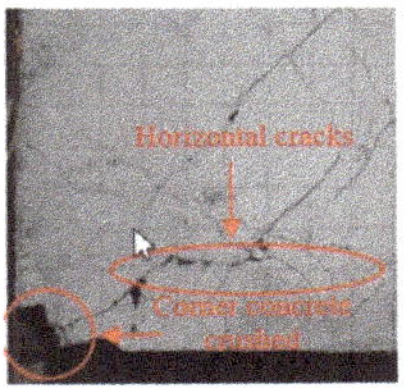

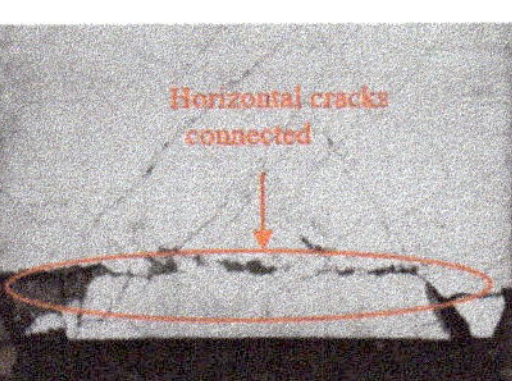

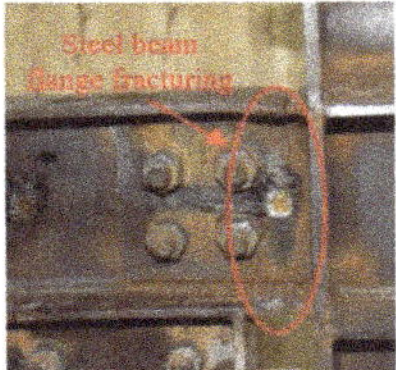

(**a**) $\delta = 1.5\delta_y$, concrete crushed   (**b**) $\delta = 2.5\delta_y$, through cracks   (**c**) $\delta = 2.5\delta_y$, relative slip   (**d**) $\delta = 5\delta_y$, beam fracturing

Figure 7. Local failure of SPE4.

(2) Specimen SPE5

Specimen SPE5 was in the elastic stage with no observable behavior when the horizontal load was less than 60 kN. Multiple fine oblique cracks appeared along the diagonal direction of the wall when the load was 80 kN. The number of fine cracks gradually increased and continuously expanded when the load was 120 kN. The specimen yielded locally, then the load was applied by controlling the displacement.

In the displacement control stage, intersecting cracks formed along the diagonal direction of the wall when the displacement was 2.5 δ_y, and the principal cracks were connected (Figure 8a). The concrete at the lower right corner of the wall was crushed and fell off when the displacement was 3.5 δ_y, exposing the steel bars. Multiple horizontal cracks with a width of 2–3 mm appeared at the bottom of the embedded T-shape connector and extended from right to left (Figure 8b). The lower left flange of the steel beam noticeably bulged upward when the displacement was 4.0 δ_y. The upper ear plate had a relative slip of approximately 10 mm with the T-shape connector, accompanied by a friction sound. A large area of concrete fell off when the displacement was 6.0 δ_y. Consequently, the steel bars were exposed, and cracks in the horizontal direction formed at the bottom of the wall (Figure 8c). The end plate warped (Figure 8d), and the column base buckled. Finally, the horizontal load decreased by more than 15%, and the specimen lost its carrying capacity.

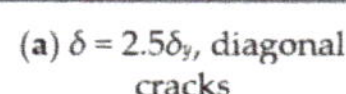

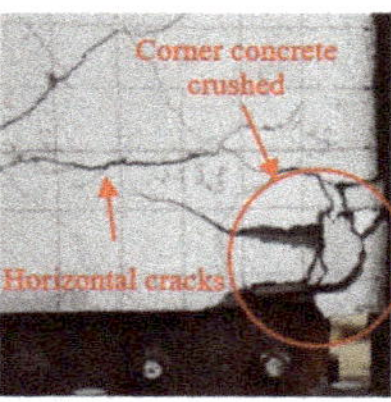

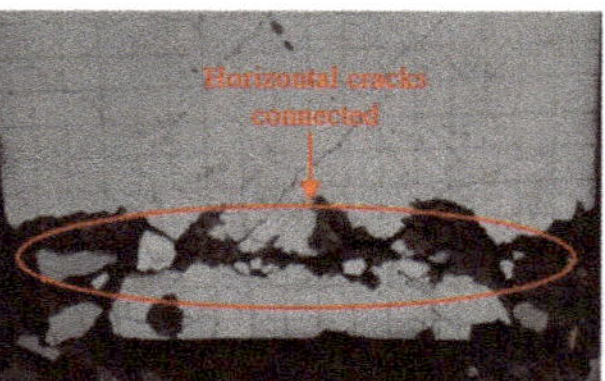

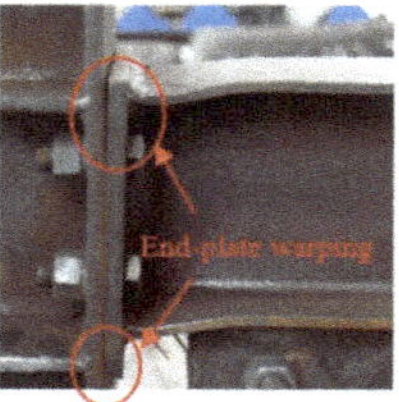

(**a**) δ = 2.5δ_y, diagonal cracks   (**b**) δ = 3.5δ_y, concrete crushed   (**c**) δ = 6δ_y, through cracks   (**d**) δ = 6δ_y, end-plate warping

Figure 8. Local failure of SPE5.

3.1.3. Pure Steel Frame

The specimen SPE6 was in the elastic stage with no observable behavior when the horizontal load was less than 140 kN. The surface coating of the steel column webs fell off locally when the load was 160 kN. A slight bending occurred at the upper flanges of the two columns when the load was 180 kN, and the test entered the displacement control loading stage. The column bases yielded when the displacement was 3.0 δ_y, and a slight out-of-plane instability occurred in the specimen. The flanges at the top portion of the left column and ends of the beam yielded when the displacement was 4.0 δ_y. The structural capacity of the specimen constantly declined and was eventually lost [35].

Several key load points according to the test behaviors of specimens SPE1 to SPE6 are summarized in Table 3.

Table 3. Several key load points in the test process.

Test Process	Specimen SPE1	Specimen SPE2	Specimen SPE3	Specimen SPE4	Specimen SPE5	Specimen SPE6
Elastic stage	Horizontal load P < 200 kN	Horizontal load P < 140 kN	Horizontal load P <180 kN	Horizontal load P < 120 kN	Horizontal load P < 60 kN	Horizontal load P < 140 kN
Cracks appeared	P = 220 kN	P = 160 kN	P = 200 kN	P = 140 kN	P = 80 kN	-
Control loading change point	P = 260 kN	P = 280 kN	P = 280 kN	P = 220 kN	P = 120 kN	P = 180 kN
Loading end point	Displacement was 6.5 δ_y	Displacement was 5.5 δ_y	Displacement was 5.5 δ_y	Displacement was 5.0 δ_y	Displacement was 6.0 δ_y	Displacement was 4.0 δ_y

3.2. Failure Modes

From the behavior of the test specimens, the force process of the specimens can be divided into four stages: elastic, concrete cracking, yield, and damage stages. In the elastic stage, steel frames and infilled RACSWs combine to resist exterior loads. The initial stiffness of the structure was high, and the load-displacement curves are linear with no observable behavior.

3.2.1. Cast-in-Place RACSWs

The horizontal load applied to the specimens of infilled cast-in-place RACSWs was transferred to the walls by the shear studs. In the concrete cracking and yield stages, initial cracking occurred along the diagonal direction of the walls, and the stiffness of the specimens decreased slightly after the cracking of the walls. The wall cracks gradually expanded and connected with the increase in horizontal load, finally forming three principal cracks with a width of 4–5 mm along the diagonal direction. The local concrete at the intersection of the principal diagonal cracks was crushed and fell off, and the energy was dissipated mainly by the coarse aggregate friction and bite of the cracked surface. A slight bulging deformation emerged at the column base and the end of the steel beam. Upon reaching the peak load, the specimens were in a damaged stage, and a large area of concrete fell off on both sides of the principal cracks. The steel bars were exposed, and the local wall showed holes. The column base and beam end buckled. The bearing capacity and lateral stiffness of the specimens decreased sharply, and the failure modes are illustrated in Figure 9a.

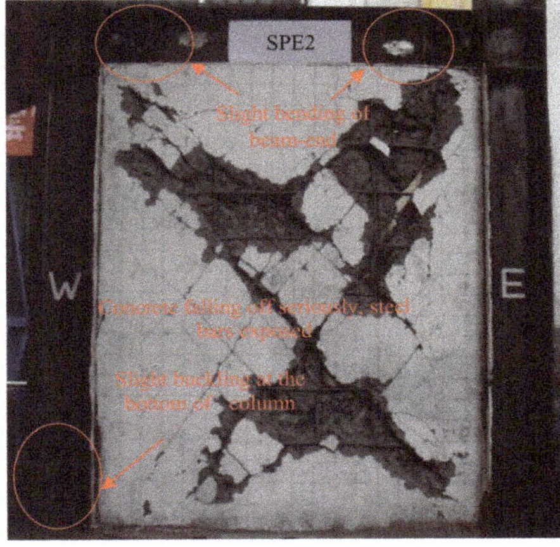

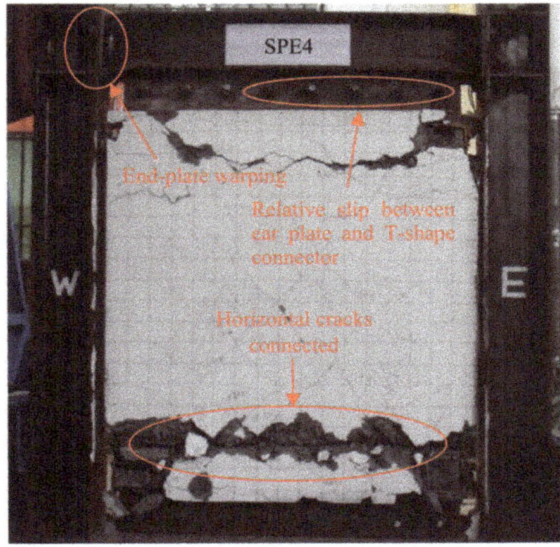

(**a**) Infilled cast-in-place wall   (**b**) Infilled prefabricated wall

Figure 9. Failure modes.

3.2.2. Prefabricated RACSWs

The horizontal load applied to the specimens of infilled prefabricated RACSWs was transferred to the walls through the ear plates, T-shape connectors, and bolts. In the concrete cracking and yield stages, initial cracking occurred along the diagonal direction of the walls. Horizontal cracks formed at the bottom of the embedded T-shape connectors because of the horizontal shear force. With an increase in horizontal load, the horizontal cracks at the embedded T-shape connectors continued to extend, and principal cracks formed with a width of 3–5 mm. The corner concrete was crushed and began to fall off. The ear plates had a relative slip with the T-shape connectors. After the peak load, a significant amount of concrete fell off, thereby exposing the steel bars. Cracks in the horizontal direction formed at the bottom of the wall. The end plates of the semi-rigid joints warped, and the flange of the steel beam of the rigid joint fractured. The column base buckled. The bearing capacity and lateral stiffness of the specimens degraded rapidly. The failure modes are depicted in Figure 9b.

3.3. Transfer Mechanism of Lateral Force

The failure modes of the specimens indicated that the horizontal load of the SFIRACSWs was resisted by the combined steel frames and infilled RACSWs.

3.3.1. Cast-in-Place RACSWs

The transfer mechanism of the lateral load of steel frames with infilled cast-in-place RACSWs is illustrated in Figure 10a. In the initial loading stage, the horizontal load was mainly resisted by the compressive strips of the wall along the diagonal direction under the effect of the extrusion pressure of the steel frame and the horizontal shear force transferred by the shear studs. With the increase in horizontal load, the wall was divided into multiple diagonal compressive strips. Then, the concrete at the compressive strips was gradually crushed. The wall gradually failed. The horizontal load was then mostly borne by the steel frame, and the bearing capacity and lateral stiffness of the specimens decreased sharply. The structure of the infilled cast-in-place RACSWs satisfied the requirements of double seismic fortification.

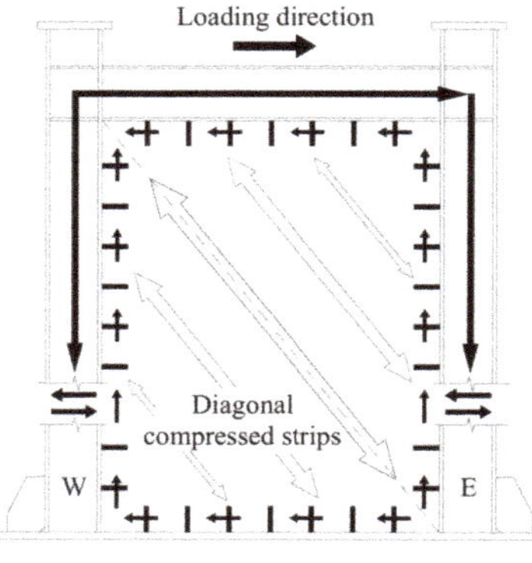

(**a**) Infilled cast-in-place wall

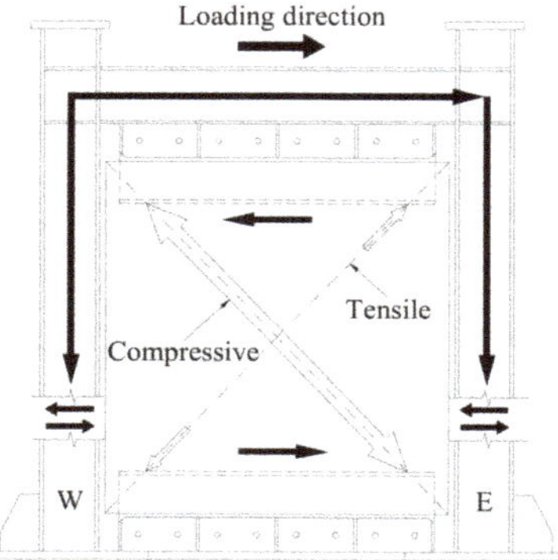

(**b**) Infilled prefabricated wall

Figure 10. Transfer mechanism of the lateral load.

3.3.2. Prefabricated RACSWs

The transfer mechanism of the lateral load of steel frames with infilled prefabricated RACSWs is illustrated in Figure 10b. In the early loading stage, the horizontal load was transferred to the wall by the ear plates, T-shape connectors, and bolts, and the wall mainly bore the horizontal shear force. With the increase in horizontal load, the wall also bore oblique compression and tension along the diagonal direction, in addition to the horizontal shear force. The wall began to crack when the stress reached the tensile strength of concrete. In the later loading stage, the wall was divided into multiple diagonal compressive strips. After the cracks in the horizontal direction formed at the bottom of the embedded T-shape connector, a large area of concrete fell off, and the wall gradually failed. The horizontal load was then mainly borne by the steel frame, and the bearing capacity and lateral stiffness of the specimens degraded rapidly.

4. Results and Discussions

4.1. Hysteretic Curves

The load-displacement hysteretic curves of the specimens are presented in Figure 11.

4.1.1. Cast-in-Place RACSWs

Figure 11a–c demonstrate the following:

(1) The specimens (Figure 11a–c) are in the elastic stage at the initial loading stage, the hysteretic curves are linear, and the loops are narrow. The hysteresis loops become spindle-shaped, and the loops open gradually with the expansion and connection of cracks in the cast-in-place RACSWs. A significant "pinch effect" occurs at the zero point. The hysteretic curves become fully arched and have a reverse S shape after the peak load because of the large area of concrete falling off in diagonal compressive strips and the plastic deformations of the steel beam and columns. The areas enclosed by the loops increase. The bearing capacity of the specimens decreases noticeably under the same load.

(2) Comparison of the hysteretic curves of the SPE1 infilled ordinary concrete wall (Figure 11a) and SPE2 infilled RACSWs with a 100% replacement rate of recycled coarse aggregate (Figure 11b) show that the area and shape surrounded by hysteretic curves and peak loads of the structure are close, thereby indicating that the performance of RACSWs is close to that of the ordinary concrete wall.

(3) The hysteretic curves of specimens SPE2 (Figure 11b) and SPE3 (Figure 11c) are relatively close, thereby showing that the connecting stiffness of BCJs slightly influences the hysteretic behavior of the specimens of infilled cast-in-place RACSWs.

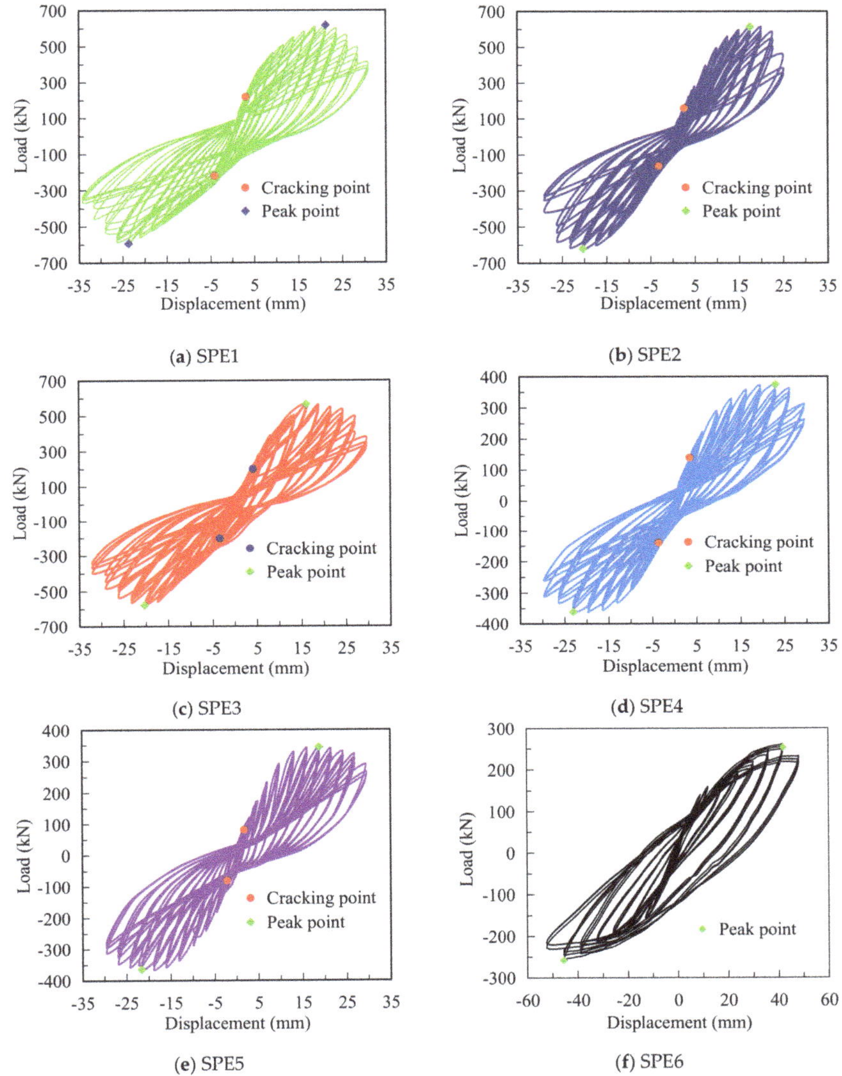

Figure 11. Hysteretic curves.

4.1.2. Prefabricated RACSWs

Figure 11d,e depict the following:

(1) The stiffness of the specimens (Figure 11d,e) is high at the initial loading stage, and the hysteretic curves are linear. Moreover, no residual deformation occurs after unloading. With the expansion and connection of cracks in the prefabricated RACSWs, the stiffness of the specimens starts to decline and the loops open gradually. Next, a significant "pinch effect" occurs at the zero point. The energy dissipation capacity of the structure is increased, the enclosed areas of the loops increase, and the hysteresis loops are spindle-shaped because of the crushing and collapse of the corner concrete on the wall, the local buckling of the steel frame, and the relative slip between the connectors. Residual deformation occurs after unloading. The hysteretic curves become fully arched after reaching the peak

load. The bearing capacity of the specimens decreases under the same load due to the large relative slip among the connectors.

(2) The hysteresis curves of SPE4 (Figure 11d) and SPE5 (Figure 11e) are nearly coincidental, indicating that the connecting stiffness of BCJs has an insignificant effect on the hysteretic behavior of the specimens of infilled prefabricated RACSWs.

4.2. Skeleton Curves

The load-displacement skeleton curves of the specimens are illustrated in Figure 12. The loads of the main characteristic points are summarized in Table 4, where P_{cr}, P_y, P_{max}, and P_u are the cracking, yield, peak, and damage loads of the specimen, respectively, and $P_u = 0.85\, P_{max}$. Figure 12 and Table 4 demonstrate the following:

(1) The skeleton curve of the pure steel frame is relatively smooth. The skeleton curves of the specimens are S-shaped when the cast-in-place RACSWs are infilled. Compared with SPE2 infilled RACSWs with a 100% replacement rate of recycled coarse aggregate, the cracking load of SPE1 infilled ordinary concrete wall increases by 37%, the average yield load decreases by 22%, and the bearing capacity is nearly the same.

(2) The bearing capacity of SPE2 is 2.4 times higher than that of the pure steel frame. The load decreases faster in SPE2 and SPE3 than in the pure steel frame after the peak load, demonstrating that the ductility of the specimens of infilled cast-in-place RACSWs decreases slightly.

(3) The comparison of SPE2 and SPE3 demonstrates that the concrete cracking load of the specimen is approximately 1.25 times higher in end-plate joints than in welded–bolted joints. The yield and peak loads decrease by 13% and 8%, respectively, showing that the connecting stiffness of BCJs slightly influences the bearing capacity of the specimens of infilled cast-in-place RACSWs.

(4) The bearing capacity is 1.44 times higher in SPE4 than in the pure steel frame when the prefabricated RACSWs are infilled, thereby indicating that the prefabricated RACSWs can effectively improve the bearing capacity of the structure.

(5) The skeleton curves of SPE4 and SPE5 are coincidental at the initial loading stage, and the peak load is only 4% lower in SPE5 than in SPE4, emphasizing that the connecting stiffness of BCJs slightly influences the bearing capacity of the specimens of infilled prefabricated RACSWs. The load of SPE4 and SPE5 decreases smoothly after the peak load, thereby indicating that the structure of the infilled prefabricated RACSWs has a high safety reserve.

Table 4. Loads of main characteristic points on skeleton curves.

Specimen	Loading Direction	Cracking Point	Yield Point	Peak Point	Failure Point
		P_{cr} (kN)	P_y (kN)	P_{max} (kN)	P_u (kN)
SPE1	Positive	219.49	402.20	618.14	525.42
	Negative	219.84	375.50	594.08	504.97
SPE2	Positive	160.15	446.50	614.32	522.17
	Negative	160.44	500.08	620.37	527.31
SPE3	Positive	200.44	437.20	566.38	481.42
	Negative	200.56	399.01	576.57	490.08
SPE4	Positive	140.12	272.60	374.62	318.43
	Negative	139.72	260.85	360.74	306.63
SPE5	Positive	80.76	265.42	343.48	291.96
	Negative	80.33	262.38	364.47	309.80
SPE6	Positive	-	157.64	252.10	225.21
	Negative	-	183.33	258.03	220.72

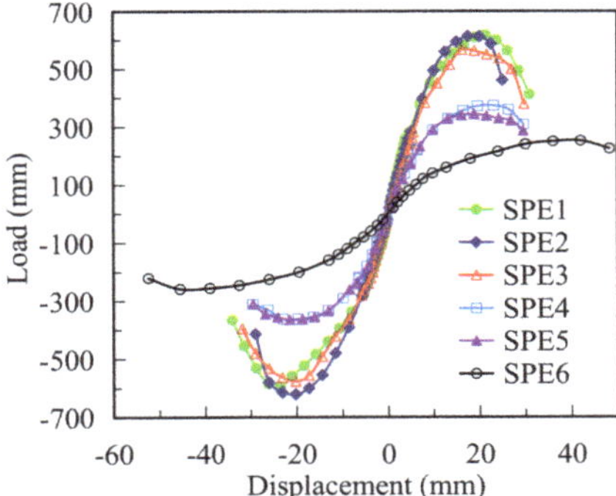

Figure 12. Skeleton curves.

4.3. Stiffness Degradation

The secant stiffness of the first cycle under the same load is calculated to reflect the degradation law of the stiffness of the specimen under cyclic loading. The formula is

$$K = \frac{|P+| + |P-|}{|\Delta+| + |\Delta-|} \tag{1}$$

where $P+$ and $P-$ are the positive and negative horizontal loads at the vertex under the same load, respectively; and $\Delta+$ and $\Delta-$ are the corresponding positive and negative horizontal displacements at the vertex under the same load.

The stiffness degradation curves of the specimens are presented in Figure 13. The values of stiffness on the main stages are provided in Table 5, where θ is the horizontal drift angle of the specimen, and K_0 is the initial stiffness of the specimen. Figure 13 and Table 5 present the following:

(1) The stiffness degradation curve of the pure steel frame is relatively smooth. The initial stiffness of SPE2 is 4.3 times higher than that of the pure steel frame when the cast-in-place RACSWs are infilled and approximately 7% lower than that of SPE1 infilled ordinary concrete wall. The comparison of SPE2 and SPE3 implies that the initial stiffness of the specimen is approximately 13% lower in end-plate joints than in welded–bolted joints, and the degradation trend of the stiffness of two specimens is basically the same.

(2) The stiffness of the specimens of infilled cast-in-place RACSWs degrades rapidly at the initial loading stage. With the increase in horizontal load, wall cracks occur and continue to expand. BCJs exhibit a slight rotation, and the stiffness degradation rate of the specimens decreases. The walls are severely damaged and gradually fail after the peak load. The drift angle of the BCJs increases, and the steel frames are used as the second seismic fortification lines to dissipate the seismic energy. The stiffness degradation of the specimens of infilled cast-in-place RACSWs stabilizes.

(3) The initial stiffness is 2.8 times higher in SPE4 than in the pure steel frame when the prefabricated RACSWs were infilled. The stiffness degradation curves of SPE4 and SPE5 are coincidental, indicating that the connecting stiffness of BCJs has an insignificant influence on the stiffness of the specimens of infilled prefabricated RACSWs. The stiffness of the specimens is degraded rapidly at the initial loading stage. The stiffness degradation rate of the specimens decreases, and the stiffness of the specimens of infilled prefabricated RACSWs is steadily reduced by expanding and connecting the cracks in the walls.

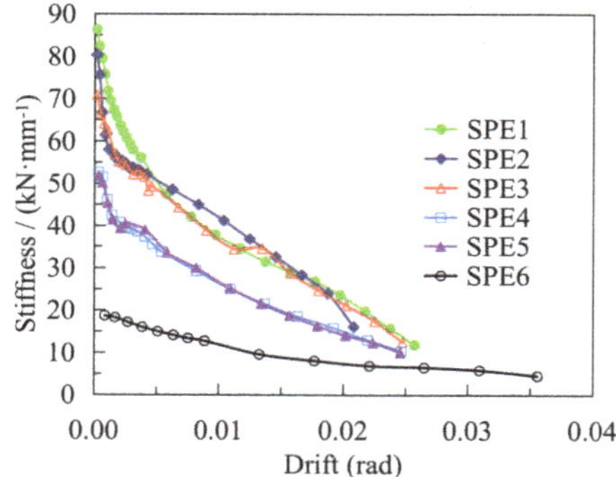

Figure 13. Stiffness degradation curves.

Table 5. Values of stiffness on main stages.

Specimen	K_0 (kN·mm^{-1})	$\theta = 0.001$ rad K_1 (kN·mm^{-1})	$\theta = 0.005$ rad K_2 (kN·mm^{-1})	$\theta = 0.010$ rad K_3 (kN·mm^{-1})	$\theta = 0.015$ rad K_4 (kN·mm^{-1})	$\theta = 0.020$ rad K_5 (kN·mm^{-1})	$\theta = 0.025$ rad K_6 (kN·mm^{-1})
SPE1	86.36	73.38	50.73	37.53	30.07	23.21	13.42
SPE2	80.39	59.91	50.77	42.02	31.81	19.34	-
SPE3	70.93	62.81	48.41	36.94	30.73	21.59	12.57
SPE4	52.75	47.01	34.87	26.57	20.15	15.14	10.45
SPE5	51.89	46.21	36.27	26.86	19.64	14.32	10.12
SPE6	18.72	18.65	15.07	11.93	8.99	7.47	6.66

4.4. Strength Degradation

The degradation coefficient (η) of the bearing capacity of the same displacement cycle is the ratio of the maximum loads of the last and first cycles. The degradation curves of the bearing capacity of the specimens are depicted in Figure 14. The values of η on the main stages are presented in Table 6. Figure 14 and Table 6 demonstrate the following:

(1) The steel frame is a typical flexible structure with good deformation capacity, and its strength degradation is unclear before the peak load. In SFIRACSWs, the walls act as the first seismic fortification lines that resist most of the horizontal load. The concrete at the diagonal compressive strips is gradually crushed and dropped, and the bearing capacity of the structure decreases sharply through the continuous expansion of the cracks in the wall.

(2) The degradation coefficients of the bearing capacity of SPE1, SPE2, and SPE3 are more than 0.97 when the horizontal drift angle is less than 0.01 rad. The degradation coefficients remain more than 0.80 when the horizontal drift angle is 0.02 rad, indicating that the specimens of infilled cast-in-place RACSWs also have a high safety reserve.

(3) The degradation coefficients of the bearing capacity of SPE4 and SPE5 are close when the horizontal drift angle is less than 0.02 rad. The degradation law is consistent, denoting that the damage degree of the specimens of infilled prefabricated RACSWs is the same under the same drift angle. Furthermore, the connecting stiffness of BCJs slightly influences the strength degradation of the specimens of infilled prefabricated RACSWs. The degradation coefficients of the bearing capacity of the specimens of infilled prefabricated RACSWs are more than 0.85.

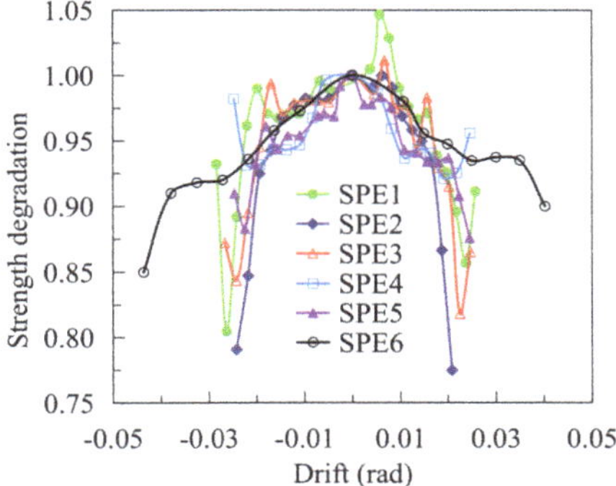

Figure 14. Strength degradation curves.

Table 6. Values of η on main stages.

Specimen	η									
	$\theta =$ −0.025 rad	$\theta =$ −0.020 rad	$\theta =$ −0.015 rad	$\theta =$ −0.010 rad	$\theta =$ −0.005 rad	$\theta =$ 0.005 rad	$\theta =$ 0.010 rad	$\theta =$ 0.015 rad	$\theta =$ 0.020 rad	$\theta =$ 0.025 rad
SPE1	0.855	0.987	0.968	0.976	0.990	1.031	0.989	0.970	0.922	0.891
SPE2	0.781	0.904	0.963	0.982	0.982	0.994	0.973	0.947	0.812	-
SPE3	0.853	0.931	0.975	0.980	0.979	0.991	0.976	0.970	0.917	0.865
SPE4	0.982	0.932	0.944	0.954	0.997	0.987	0.945	0.942	0.922	0.956
SPE5	0.910	0.935	0.947	0.958	0.970	0.981	0.954	0.936	0.937	0.876
SPE6	0.926	0.943	0.961	0.975	0.988	0.991	0.981	0.956	0.948	0.935

4.5. Ductility Analysis

Displacement ductility factor is the ratio of damage displacement Δ_u to yield displacement Δ_y, which is an important index for measuring the deformation capability of a structure. The inter-story drift angles and displacement ductility factors of the main stages are listed in Table 7, where Δ_{cr}, Δ_y, Δ_{max}, and Δ_u are the cracking, yield, peak, and damage displacements of the specimen, respectively; and θ_{cr}, θ_y, θ_{max}, and θ_u are the cracking, yield, peak, and damage drift angles, respectively. Table 7 shows the following:

(1) The displacement ductility factor of the pure steel frame is 3.47. The displacement ductility factors of the specimens are from 2.44 to 2.69 when the cast-in-place RACSWs are infilled. The infilled walls can increase the bearing capacity and initial stiffness of the structure while reducing the yield and damage displacement of the structure. Thus, the ductility of the specimens of infilled cast-in-place RACSWs is reduced.

(2) The displacement ductility coefficient in an SPE1 infilled ordinary concrete wall is 1.34 times that of SPE2, indicating that the wall made of recycled coarse aggregate has poor bonding performance and ductility.

(3) The inter-story drift angles are from 1/415 to 1/317 at the concrete cracking stage, from 1/116 to 1/114 at the yield stage, and from 1/66 to 1/64 at the peak point, thereby indicating that the specimens of infilled cast-in-place RACSWs have a good deformation capacity. The displacement ductility factor is approximately 10% higher in SPE3 than in SPE2, suggesting that the specimen of end-plate joints is simple to construct and has a good deformation capacity, and the ductility is better in the end-plate joints than in the welded–bolted joints.

Table 7. Displacement and displacement ductility factors.

Specimen	Loading Direction	Cracking Point			Yield Point			Peak Point			Failure Point			Ductility Factors	
		Δ_{cr} (mm)	θ_{cr}	Average	Δ_y (mm)	θ_y	Average	Δ_{max} (mm)	θ_{max}	Average	Δ_u (mm)	θ_u	Average	μ	Average
SPE1	Positive	3.07	1/391	1/332	7.73	1/155	1/136	21.30	1/56	1/53	27.43	1/44	1/42	3.55	3.28
	Negative	4.15	1/289		9.90	1/121		23.71	1/51		29.77	1/40		3.01	
SPE2	Positive	2.64	1/454	1/415	8.78	1/137	1/114	17.51	1/69	1/64	23.80	1/50	1/47	2.71	2.44
	Negative	3.14	1/382		12.41	1/97		20.31	1/59		27.02	1/44		2.18	
SPE3	Positive	4.15	1/289	1/317	10.30	1/117	1/116	16.20	1/74	1/66	27.41	1/44	1/43	2.66	2.69
	Negative	3.41	1/352		10.48	1/115		20.30	1/59		28.42	1/42		2.71	
SPE4	Positive	3.55	1/338	1/337	8.92	1/135	1/137	23.06	1/52	1/52	29.03	1/41	1/41	3.25	3.32
	Negative	3.56	1/337		8.59	1/140		23.02	1/52		29.11	1/41		3.39	
SPE5	Positive	1.76	1/682	1/619	8.59	1/140	1/139	18.83	1/64	1/60	29.13	1/41	1/41	3.39	3.40
	Negative	2.12	1/566		8.68	1/138		21.56	1/56		29.50	1/41		3.40	
SPE6	Positive	-	-	-	14.15	1/85	1/83	42.00	1/29	1/27	48.28	1/25	1/24	3.41	3.47
	Negative	-	-		14.88	1/81		45.51	1/26		52.41	1/23		3.52	

(4) The displacement ductility factors of the specimens are from 3.32 to 3.40 when the prefabricated RACSWs are infilled. The overall deformation is restrained by the walls, and the horizontal displacements are smaller in the main stages than those in the pure steel frame, although the bearing capacity and lateral stiffness of the specimens of infilled prefabricated RACSWs are remarkably increased. Consequently, the displacement ductility factors are slightly lower in the specimens of infilled prefabricated RACSWs than those in the pure steel frame. The inter-story drift angle is from 1/619 to 1/337 at the concrete cracking stage, from 1/139 to 1/137 at the yield stage, and from 1/60 to 1/52 at the peak point.

4.6. Energy Dissipation Capacity

The energy dissipation capacity of the specimens is expressed by the relation curves between the hysteretic loop area and the horizontal drift angle, as shown in Figure 15. The values of energy dissipation in the main stages are presented in Table 8.

(1) The energy of the specimens of infilled cast-in-place RACSWs are dissipated mainly by the flexible deformation of the steel frames and the coarse aggregate friction and bite of the cracked surface of the RACSWs. The energy dissipation is 3.25 times higher in SPE2 infilled cast-in-place RACSWs than in the pure steel frame when $\theta = 0.005$ rad, and 2.6 times higher in SPE2 than in the pure steel frame when $\theta = 0.02$ rad.

(2) The energy dissipation of SPE2 infilled RACSWs is 41% that of SPE1 infilled ordinary concrete wall when $\theta = 0.005$ rad, and 56% that of SPE1 when $\theta = 0.02$ rad. The energy dissipation is approximately 13% higher in SPE3 than in SPE2 when $\theta = 0.005$ rad, and approximately 28% higher in SPE3 than in SPE2 when $\theta = 0.02$ rad, thereby indicating that the end-plate joints are fully deformed and characterized by excellent energy dissipation during loading.

(3) The energy of the specimens of infilled prefabricated RACSWs is dissipated mainly by the flexible deformation of the steel frames and the coarse aggregate friction and bite of the wall cracks and friction slip among the connectors. In the early stage of loading, the cracks on the wall of SPE5 occur early and the concrete cracking load is low; the energy dissipation is slightly higher in SPE5 than in SPE4. With the increase in displacement, the energy dissipation of the two specimens becomes the same. The energy dissipation capacity is approximately two times higher in the specimens of infilled prefabricated RACSWs than in the pure steel frame.

(4) Compared with the pure steel frame, infilled RACSWs can greatly improve the stiffness and energy dissipation capacity of SFIRACSWs while reducing the ductility of the structure, thereby indicating that infilled RACSWs strongly influence the hysteretic behavior of SFIRACSWs.

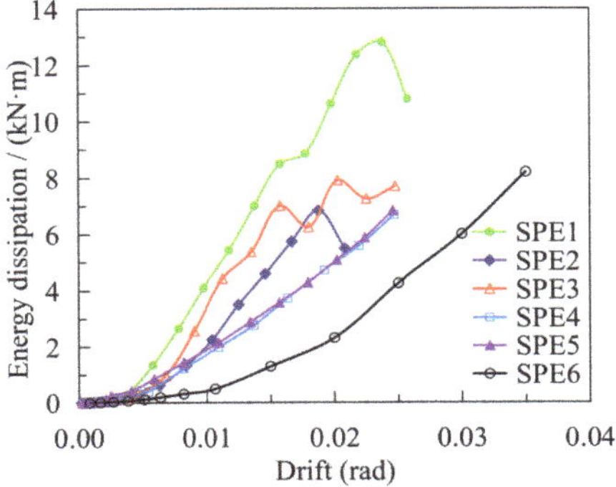

Figure 15. Energy dissipation curves.

Table 8. Values of energy dissipation at main stages.

Specimen	Energy Dissipation (kN·m)				
	$\theta = 0.005$ rad	$\theta = 0.010$ rad	$\theta = 0.015$ rad	$\theta = 0.020$ rad	$\theta = 0.025$ rad
SPE1	0.96	4.27	7.93	10.83	11.55
SPE2	0.39	2.07	4.81	6.03	-
SPE3	0.44	3.38	6.45	7.71	7.70
SPE4	0.36	1.74	3.26	4.99	6.80
SPE5	0.64	1.9	3.36	5.01	7.01
SPE6	0.12	0.45	1.30	2.32	4.25

5. Conclusions

The following conclusions can be drawn from the low cyclic experiments on steel frames with infilled cast-in-place RACSWs and prefabricated RACSWs:

(1) The bearing capacity and initial stiffness were 2.4 and 4.3 times higher in the steel frames with infilled cast-in-place RACSWs than those in the pure steel frame. The displacement ductility factors were from 2.44 to 2.69. The degradation coefficients of the bearing capacity remained over 0.80 when the horizontal drift angle was 0.02 rad, thereby indicating that the specimens of infilled cast-in-place RACSWs had a high safety reserve.

(2) Compared with the cracking load of the specimen of infilled RACSWs with a 100% replacement rate of recycled coarse aggregate, that of the infilled ordinary concrete wall increased by 37%, the yield load decreased by 22%, and the bearing capacity was nearly the same. These results indicate that the performance of RACSWs was nearly the same as that of an ordinary concrete wall in the structure of steel frames with infilled shear walls.

(3) The yield and peak loads of the specimen decreased by only 13% and 8%, respectively, in the end-plate joints compared with those in the welded–bolted joints. Furthermore, the initial stiffness was reduced by approximately 13%. The infilled cast-in-place RACSWs relieved the rotation deformation of semi-rigid joints and weakened the influence of the connecting stiffness of BCJs on the bearing capacity of the structure of infilled cast-in-place RACSWs.

(4) The bearing capacity and initial stiffness were 1.44 and 2.8 times higher in the steel frames with infilled prefabricated RACSWs than those in the pure steel frame, and the displacement ductility factors were from 3.32 to 3.40. The difference in bearing capacity of the specimens in the welded–bolted and end-plate joints was only 4%, and the turning capability and ductility were better in the semi-rigid joints than in the rigid joints.

(5) The connectors between the steel frames and prefabricated RACSWs were undamaged during the test, and the shear force was transferred successfully. The cracks in the horizontal direction were formed at the connection between the embedded T-shape connectors and the walls, and shear failure occurred in the specimens. Therefore, the connection construction between the embedded T-shape connectors and walls should be given sufficient attention.

(6) The prefabricated shear walls made of recycled coarse aggregate improved the lateral stiffness and bearing capacity of the structure of infilled prefabricated RACSWs. The structure of infilled prefabricated RACSWs was characterized by a favorable deformation capability to satisfy the design requirements of structure behavior in the seismic fortification area. The walls and steel frames could be rapidly installed in the construction field. Furthermore, the structure of the infilled prefabricated RACSWs was safe, highly efficient, convenient to repair and replace after an earthquake, and had a satisfactory engineering application value.

Author Contributions: Conceptualization, L.S. and H.G.; methodology, Y.L.; software, Y.L.; validation, L.S., H.G.; and Y.L.; formal analysis, L.S.; investigation, L.S.and H.G.; resources, H.G.; data curation, L.S.; writing—original draft preparation, L.S.; writing—review and editing, L.S. and H.G.; visualization, L.S.; supervision, Y.L.; project administration, H.G.; funding acquisition, H.G. and Y.L.

Funding: This research was funded by the National Natural Science Foundation of China No. 51308454.

Conflicts of Interest: The authors declare no conflicts of interest.

References

1. Puthussery, J.V.; Kumar, R.; Garg, A. Evaluation of recycled concrete aggregates for their suitability in construction activities: An experimental study. *Waste Manag.* **2016**, *60*, 270–276.
2. Tabsh, S.W.; Abdelfatah, A.S. Influence of recycled concrete aggregates on strength properties of concrete. *Constr. Build. Mater.* **2009**, *23*, 1163–1167.
3. Koenders, E.A.B.; Pepe, M.; Martinelli, E. Compressive strength and hydration processes of concrete with recycled aggregates. *Cem. Concr. Res.* **2014**, *56*, 203–212.
4. Silva, R.V.; Brito, J.D.; Dhir, R.K. Tensile strength behaviour of recycled aggregate concrete. *Constr. Build. Mater.* **2018**, *83*, 108–118.
5. Bairagi, N.K.; Ravande, K.; Pareek, V.K. Behaviour of concrete with different proportions of natural and recycled aggregates. *Resour. Conserv. Recycl.* **1993**, *9*, 109–126.
6. Oikonomou, N. Recycled concrete aggregates. *Cem. Concr. Compos.* **2005**, *27*, 315–318.
7. Ying, J.; Xiao, J.; Tam, V.W.Y. On the variability of chloride diffusion in modelled recycled aggregate concrete. *Constr. Build. Mater.* **2013**, *41*, 732–741.
8. Wang, C.; Xiao, J.; Zhang, G.; Li, L. Interfacial properties of modeled recycled aggregate concrete modified by carbonation. *Constr. Build. Mater.* **2016**, *105*, 307–320.
9. Arezoumandi, M.; Smith, A.; Volz, J.S.; Khayat, K.H. An experimental study on shear strength of reinforced concrete beams with 100% recycled concrete aggregate. *Constr. Build. Mater.* **2014**, *53*, 612–620.
10. Choi, W.C.; Yun, H.D. Long-term deflection and flexural behavior of reinforced concrete beams with recycled aggregate. *Mater. Des.* **2013**, *51*, 742–750.
11. Choi, W.C.; Yun, H.D. Compressive behavior of reinforced concrete columns with recycled aggregate under uniaxial loading. *Eng. Struct.* **2012**, *41*, 285–293.
12. Xiao, J.; Huang, X.; Shen, L. Seismic behavior of semi-precast column with recycled aggregate concrete. *Constr. Build. Mater.* **2012**, *35*, 988–1001.
13. Wu, B.; Zhang, S.; Yang, Y. Compressive behaviors of cubes and cylinders made of normal-strength demolished concrete blocks and high-strength fresh concrete. *Constr. Build. Mater.* **2015**, *78*, 342–353.
14. Wu, B.; Zhao, X.Y.; Zhang, J.S. Cyclic behavior of thin-walled square steel tubular columns filled with demolished concrete lumps and fresh concrete. *J. Constr. Steel Res.* **2012**, *77*, 69–81.
15. Chen, Z.; Jing, C.; Xu, J.; Zhang, X. Seismic performance of recycled concrete-filled square steel tube columns. *Earthq. Eng. Eng. Vib.* **2017**, *16*, 119–130.
16. Dong, J.F.; Wang, Q.Y.; Guan, Z.W. Structural behaviour of recycled aggregate concrete filled steel tube columns strengthened by CFRP. *Eng. Struct.* **2013**, *48*, 532–542.
17. Fathifazl, G.; Razaqpur, A.G.; Isgor, O.B.; Abbas, A.; Fournier, B.; Foo, S. Shear capacity evaluation of steel reinforced recycled concrete (RRC) beams. *Eng. Struct.* **2011**, *33*, 1025–1033.
18. Ma, H.; Xue, J.; Zhang, X.; Luo, D. Seismic performance of steel-reinforced recycled concrete columns under low cyclic loads. *Constr. Build. Mater.* **2013**, *48*, 229–237.
19. Corinaldesi, V.; Letelier, V.; Moriconi, G. Behaviour of beam-column joints made of recycled-aggregate concrete under cyclic loading. *Constr. Build. Mater.* **2011**, *25*, 1877–1882.
20. Gonzalez, V.C.L.; Moriconi, G. The influence of recycled concrete aggregates on the behavior of beam-column joints under cyclic loading. *Eng. Struct.* **2014**, *60*, 148–154.
21. Xiao, J.; Wang, C.; Li, J.; Tawana, M.M. Shake-table model tests on recycled aggregate concrete frame structure. *ACI Struct. J.* **2012**, *109*, 777–786.
22. Wang, C.; Xiao, J. Study of the seismic response of a recycled aggregate concrete frame structure. *Earthq. Eng. Eng. Vib.* **2013**, *12*, 669–680.
23. Peng, Y.; Wu, H.; Yan, Z. Strength and drift capacity of squat recycled concrete shear walls under cyclic loading. *Eng. Struct.* **2015**, *100*, 356–368.
24. Ma, G.; Zhang, Y.; Liu, Y.; Li, Z. Seismic behaviour of recycled aggregate thermal insulation concrete (Ratic) shear walls. *Mag. Concr. Res.* **2015**, *67*, 145–162.

25. Zhang, J.; Cao, W.; Meng, S.; Yu, C.; Dong, H. Shaking table experimental study of recycled concrete frame-shear wall structures. *Earthq. Eng. Eng. Vib.* **2014**, *13*, 257–267.

26. Tong, X.; Hajjar, J.F.; Schultz, A.E.; Shield, C.K. Cyclic behavior of steel frame structures with composite reinforced concrete infill walls and partially-restrained connections. *J. Constr. Steel Res.* **2005**, *61*, 531–552.

27. Sun, G.; He, R.; Qiang, G.; Fang, Y. Cyclic behavior of partially-restrained steel frame with RC infill walls. *J. Constr. Steel Res.* **2011**, *67*, 1821–1834.

28. Kurata, M.; Leon, R.T.; DesRoches, R.; Nakashima, M. Steel plate shear wall with tension-bracing for seismic rehabilitation of steel frames. *J. Constr. Steel Res.* **2011**, *71*, 92–103.

29. Guo, H.C.; Hao, J.P.; Liu, Y.H. Behavior of stiffened and unstiffened steel plate shear walls considering joint properties. *Thin-Walled Struct.* **2015**, *97*, 53–62.

30. Guo, H.C.; Li, Y.L.; Liang, G.; Liu, Y.H. Experimental study of cross stiffened steel plate shear wall with semi-rigid connected frame. *J. Constr. Steel Res.* **2017**, *135*, 69–82.

31. Wu, H.H.; Zhou, T.H.; Liao, F.F.; Lv, J. Seismic behavior of steel frames with replaceable reinforced concrete wall panels. *Steel Compos. Struct.* **2016**, *22*, 1055–1071.

32. Wang, J.; Li, B. Cyclic testing of square CFST frames with ALC panel or block walls. *J. Constr. Steel Res.* **2017**, *130*, 264–279.

33. Hou, H.; Chou, C.C.; Zhou, J.; Wu, M.; Qu, B.; He, H.; Liu, H.; Li, J. Cyclic tests of steel frames with composite lightweight infill walls. *Earthq. Struct.* **2016**, *10*, 163–178.

34. Geng, Y.; Wang, Y.; Ding, J.; Xu, W. Mechanical behavior of connections between out-hung light-gauge steel stud walls and steel frames. *J. Build. Struct.* **2016**, *37*, 141–150. (In Chinese)

35. Sun, L.J.; Guo, H.C.; Liu, Y.H. Study on seismic behavior of steel frame with external hanging concrete walls containing recycled aggregates. *Constr. Build. Mater.* **2017**, *157*, 790–808.

© 2019 by the authors. Licensee MDPI, Basel, Switzerland. This article is an open access article distributed under the terms and conditions of the Creative Commons Attribution (CC BY) license (http://creativecommons.org/licenses/by/4.0/).

Article

An Experimental Strain-Based Study on the Working State of Husk Mortar Wallboards with Openings

Xuesong Cai [1], Shijun Sun [2,*] and Guangchun Zhou [3]

[1] Department of Geotechnical Engineering, Tongji University, Shanghai 200092, China; caixiaoshuai@tongji.edu.cn

[2] Key Laboratory of Ministry of Industry and Information Technology for Human Settlement Environmental Science and Technology School of Architecture, Harbin Institute of Technology, Harbin 150090, China

[3] Key Lab of Smart Prevention and Mitigation of Civil Engineering Disasters of the Ministry of Industry and Information Technology, Harbin Institute of Technology, Harbin 150090, China; gzhou@hit.edu.cn

* Correspondence: sunsj@hit.edu.cn

Received: 29 November 2019; Accepted: 17 January 2020; Published: 19 January 2020

Abstract: Rice husks as common agricultural remnants with low density and good thermal conductivity properties have been used in infill walls in the northern area of China. Accordingly, many tests and numerical simulations were conducted to address a difficult issue, the inaccurate estimation on the lateral load-bearing capacity of different types of husk mortar energy-saving (HMES) wallboards. The difficulty has not been overcome so far, implying that the novel methods are anticipated to achieve the accurate estimation. This paper tests the full-scale HMES wallboards with different openings and obtains the strains at the points distributed on the wallboard sides. The experimental strains are modeled as the approximate strain energy values to produce the characteristic parameter of the HMES wallboard's stressing state. Furthermore, the inherent working state characteristic points of HMES wallboards are revealed from the evolution of the characteristic parameter called as the normalized approximate strain energy sum, leading to the redefinition of the failure loads for the HMES wallboards. Finally, it investigates the stressing state mode evolution of the HMES wallboard around the failure loads. The achieved results provide the reference to the accurate estimation of the bearing capacity of the HMES wallboards.

Keywords: husk mortar wallboard; experiment; lateral strength; strain; failure load

1. Introduction

To address the global issues of the warming climate and pollution, many countries have set "energy conservation" and "environmental protection" goals in recent decades. Meanwhile, many countermeasures have been taken based on the individual national conditions [1]. For instance, China banned or limited construction materials that contaminate the environment, such as clay bricks and natural woods [2], and encouraged energy-saving and environmentally friendly construction materials through policies and subsidies [3]. As a result, many new types of structural components have been developed, inclusive of rice husk and wheat husk wall panels. Commonly, energy-saving construction components are composite so that their working behavior or working mechanism are complicated and their load-bearing capacity is difficult to be accurately estimated, particularly, for their seismic performance [4,5]. The following makes an overview of the relative research results.

In the respect of experimental investigations, monotonic loading tests, quasi-static tests, and shaking table tests on different types of infill walls have been performed in recent years [6–8]. Although the tests could verify the analytical results of the proposed methods, they also presented the great variation of infill walls in configuration, material property, and manufacture of composite components. Likewise, Peng et al. [9] conducted three cyclic experiments of infill walls with a UHPC (Ultra high performance concrete) layer and a reinforced polymer mortar layer. The UHPC layer improved a typical historical squat wall's resistance by 193%, cracking load by 127%, and ultimate deformation by 109%. Dong et al. [10] used reinforced mortar cross strips to strengthen the masonry walls and the quasi-static test results showed that such strengthening method was an effective way to improve the shear capacity and seismic performance of masonry walls. Similarly, pre-stressing technology was used by Chi et al. [11] in fully grouted concrete masonry wall systems to promote its seismic performance. Through the parameters including displacement ductility, stiffness degradation, energy dissipation, and equivalent viscous damping, the test results verified the advantages of the pre-stressing technology. Moreover, from the shaking table test of a 5-story prototype structure with a reinforced concrete frame and unreinforced masonry, Hashemi et al. [12] found that the unreinforced masonry infill wall provided a significant contribution to the strength and ductility of the structure and should be considered in both analysis and design.

In the respect of numerical modelling, finite element (FE) analysis was widely used to study the mechanical properties of infill walls [13–16]. Betti et al. [17] built both plasticity model and smeared cracking/crushing model to analyze the working behavior of historic buildings with infill walls under seismic loading. Bartoli et al. [18] suggested the multi-level approach to assess the static and seismic capacity of historic masonry towers, and performed three levels of evaluation, namely LV1 (analysis at territorial level), LV2 (local analysis), and LV3 (global analysis). Ferrero et al. [19] calibrated the numerical model according to the eigenvalue analysis and the test of the dynamic identification. Then the nonlinear static analyses were performed to evaluate the seismic response of the structure, which obtained the reliable parameters for modeling infill walls. In the recent years, Zhou et al. [20] built the neural network mode of laterally loaded masonry walls to estimate their failure loads and failure patterns. Zhang et al. [21] used the cellular automata approaches to predict the cracking patterns of masonry wallets under vertical loads. Huang et al. [22] developed a generalized strain energy density (GSED)–based method and revealed the relationship between the failure loads of the base/test masonry wall panels.

The available data regarding full-scale wallboard's load capacity remains limited. Investigations on the strain data for analyzing structural behavior are particularly scarce. Most of these researches focus on the load and displacement data. No experimental program has yet, to the authors' knowledge, been undertaken with strain analysis method on husk mortar energy-saving (HMES) wallboards. The tests and numerical analyses mentioned above obviously reflect two issues: (1) A reasonable numerical model would not have been established due to the configuration complexity of HMES wallboards. As a result, an accurate prediction for the ultimate strength of HMES wallboards has not been achieved by the existing analytical theories and methods. (2) The displacement data and force data were fully utilized, but the strain data were not fully applied. The structural working features embodied in the strain data have not been explored. Both problems implied that new theories and methods should be developed to solve them [23]. Zhou's structural stressing state theory explored a new way to deeply understand the structural working behavior features from the experimental strain data and provided a proper method to definite structural failure load [24,25]. In view of Zhou's theory, this paper investigated the structural working state of full-scale HMES wallboards with different openings. It was considered that the structural working state related to the individual working states of the local zones within the wallboard. So the strain values with the load increase were measured at the points uniformly distributed on both sides of the wallboard. The study revealed the inherent leap characteristics implied in the working states of wallboards and redefined their failure loads. An empirical relationship among the experimental failure loads of the wallboards was derived based on the failure characteristics of the wallboards. which could provide the reference to the accurate prediction of the wallboards' load-bearing capacity. In addition, a new way was explored to apply the enormous amount of experimental strain data to structural analysis.

2. HMES Wallboards and Testing Preparation

2.1. The Configuration and Size of the HMES Wallboards

The structure of HMES wallboards is briefly described since it has already been reported in detail [26]. The husk mortar is made of cement, sand, husks, water, and additive. This panel can be used to the multi-story and high-rise reinforced concrete frame construction as facade, its overall size is the same with the external wall. The wall panel consists of two husk mortar surface layers, and two insulation layers (benzene or asbestos), and an air layer inside. The two surface layers are connected by the vertical and horizontal reinforced ribs.

Figure 1 shows the sectional structure. An HMES wallboard consists of three parts: two surface layers, two benzene board layers, and one cavity layer. The vertical and transverse ribs are arranged in the interior of the wallboard, and the reinforcing bars are arranged between the two surfaces. The metal mesh is in the surface layer. The benzene board is cut into small pieces, and between two pieces, there is a gap for a rib. The insulation mortar is poured between the surface layer and the benzene board layer for bonding and fixation. The vertical reinforced bar, the metal mesh in the surface layer, and the transverse pull bar are bonded at their intersection, constituting a three-dimensional metal grid. The metal rings are set in the upper edge of the wallboard for convenient hoisting and transport. Figure 2 shows the wallboard products.

Figure 1. The wallboard products.

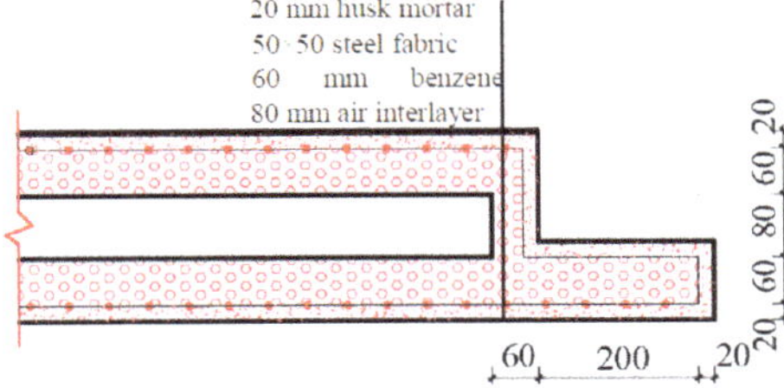

Figure 2. Section of husk mortar energy-saving (HMES) wallboard.

As shown in Figure 3, an experimental building with the HMES wallboards was built in 2008, and some houses were built later with such wallboards according to needs of local construction. The dimensions of HMES wallboards are determined by column space, height of the floor, and dimensions of windows and doors.

In this experiment, two groups of full-scale HMES wallboards are manufactured based on the norms of practical engineering. The sizes, lateral loading cases, and numbers of the wallboards are shown in Table 1. Considering their low lateral bearing capacity, which is unsuitable for cyclic loading testing, wallboards 1-3, 1-4, 2-1, 2-2, 2-3, and 2-4 with openings are subjected to monotonic loads only. For purposes of comparison with the monotonic loading cases, wallboards 1-1 and 1-2 are subjected to cyclic loads.

Figure 3. Construction process.

Table 1. The configurations and sizes of the experimental HMES wallboards.

Configuration & Size Dimensions in mm	Group 1		Group 2	
	No.	Lateral Loading Case	No.	Lateral Loading Case
Intact Wallboard 3300 × 2800 × 250	1-1	cyclic	2-1	monotonic
Window opening: 1500 × 1600 3300 × 2800 × 250	1-2	cyclic	2-2	monotonic
Door opening: 900 × 2100 2400 × 2800 × 250	1-3	monotonic	2-3	monotonic
Window opening:1500 × 1600 Door opening: 900 × 2100 3300 × 2800 × 250	1-4	monotonic	2-4	monotonic

2.2. Experimental Device and Loading Scheme

This experiment was carried out at the Heilongjiang Institute of Science and Technology, China. The experimental device is shown in Figure 4. The vertical load was applied by mechanical jacks, and the horizontal load was applied by a hydraulic servo actuator, which was installed in the same plane as the specimen. To facilitate lifting, a steel groove was designed and fixed at the bottom of the tested wallboard by pouring concrete. Two jacks were installed for connecting the specimen and the base. In this way, the base of the specimen was fixed during the test. A pair of two tie rods were installed on the loading end of the horizontal actuator; a cyclic load could be realized with such a loading device. The horizontal cyclic load and monotonic loads were supplied by a 500 kN actuator. As shown in Figure 5, the loading was force controlled for the cyclic loading case as well as the monotonic loading case. The loading scheme was composed of applying incremental multiples of 5.0 kN until the test was terminated for the specimen 1-1, specimen 1-2, specimen 2-1, and specimen 2-2. As for specimen 1-3, specimen 1-4, specimen 2-3, and specimen 2-4, the incremental force was 3.0 kN. In addition, the specimen was exposed to an axial load of 30 kN at the top surface that was applied by two hydraulic jacks on the rigid distribution beam, which represented the gravity load and remained constant throughout the test. The force controlled scheme was chosen over displacement controlled scheme for the reason that force and strain energy data were significant in the analysis of strain energy, which is applied in Section 4.3. As the load capacities of the HMES wallboards with different sizes are usually different, the load increments of specimen 1-1 and 1-2 are set 5.0 kN while the load increments of specimen 1-3 and 1-4 are set 3.0 kN.

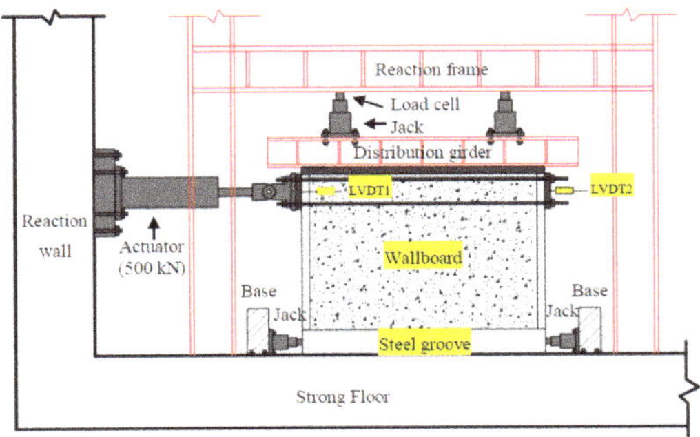

Figure 4. Overview of the test setup and instrumentation.

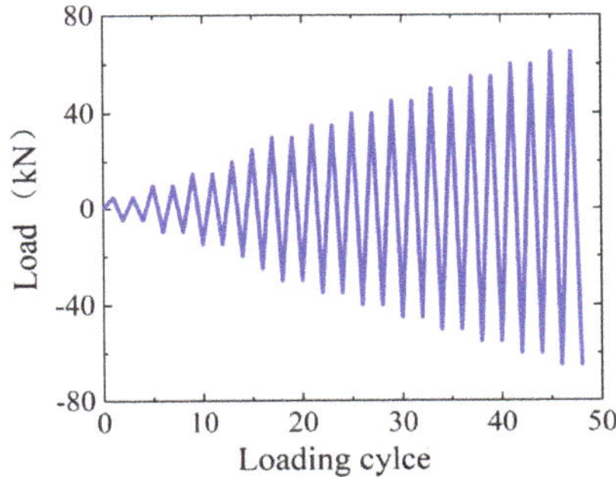

Figure 5. The loading scheme for the cycle test.

2.3. Measuring Scheme

During the HMES wallboard experiment, the measuring data include the horizontal load values and the corresponding displacements at the loading position. To measure strain, standard strain gauges specific to concrete are adopted. The strain signal is recorded by the electrical resistance strain indicator per second. The strain data are collected at the peak point of every loading process. Notably, this experiment specifically considers the measurement of the strain values: (1) the strain values indicate the local behavior of the wallboard; thus, they are recorded in three directions at the representative positions on both sides of the wallboard; and (2) the working state mode of the wallboard can be numerically described by the strain values at uniformly distributed locations. Hence, for instance, the strain layout of wallboard 1-1 is designed as shown in Figure 6.

Two linear variable differential transducers (LVDTs) were placed at the top sections of the specimens to measure longitudinal deformations along the loading direction. All measurements of the load cells, displacement transducers, and strain gauges were recorded by a computer data logger.

The principal strains, the shear strains, and the inclinations of the principal strains were calculated based on the strain data, and the maximum values of the strains under each level of loading were applied for calculating the strain energy.

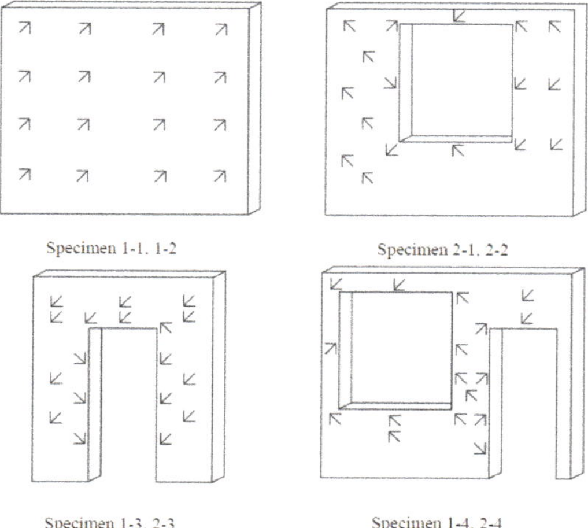

Figure 6. The layout of the strain gauges.

3. The Experimental Results

3.1. Test Results of Material Test

As shown in Figure 7, material tests of husk mortar were carried out according to the requirements of the Standard of Testing of Mechanical Properties of Ordinary Concrete [27]. The compressive strength of husk mortar is 18.6 MPa, measured on a set of 3 cubic samples. Several key parameters of husk mortar, such as compressive strength, density, water content, thermal conductivity, and elastic modulus are listed in Table 2.

Figure 7. Cube for material test.

Table 2. Mechanical properties of husk mortar.

Cube Number	Compressive Strength	Density	Water Content	Thermal Conductivity	Elastic Modulus
	MPa	kg/m³	%	W/(m × k)	MPa
A	21.20	1733	3.1	0.394	
B	14.13	1467	3.5	0.233	4976
C	20.60	1633	3.4	0.380	

3.2. Test Results of HMES Wallboords

Typically, experimental data are analyzed to have a reliable and accurate formula for calculating the ultimate/failure load and the structural parameters corresponding to some working behavior characteristics. For the experimental data related to this study, such an analysis was also performed [20]. However, it seems that the experimental records of cracking phenomena and strain values have not been fully studied thus far. A great amount of unseen knowledge is certainly implied in the experimental data, and only some innovative method is necessary to reveal this knowledge. Here, the analysis of the experimental data focuses on the cracking mechanism and the stressing state of the wallboard from a structural perspective based on the concept of the structural stress state.

Figure 8 draws the failure patterns of the eight wallboards based on the experimental observations. It should be noted that the concrete bonding failure occurs in the steel groove of specimen 2-1. Through upgrading the strength of the concrete, which was poured in the steel groove, similar failure patterns do not reappear in the following tests.

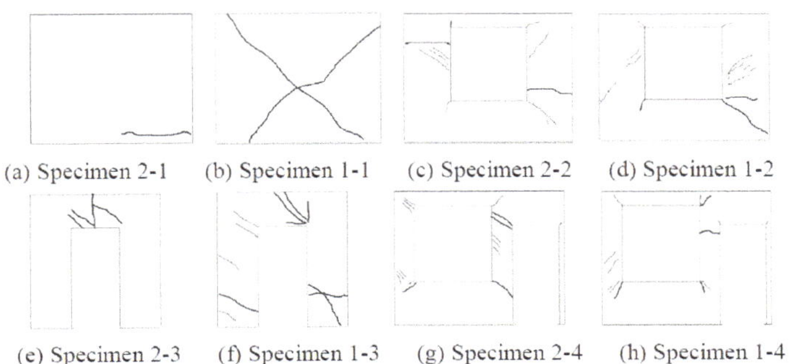

(a) Specimen 2-1   (b) Specimen 1-1   (c) Specimen 2-2   (d) Specimen 1-2

(e) Specimen 2-3   (f) Specimen 1-3   (g) Specimen 2-4   (h) Specimen 1-4

Figure 8. The experimental cracking patterns of the HMES wallboards.

The experimental records of the HMES wallboards corresponding to Table 1 are summarized in Table 3.

Table 3. The experimental records of the HMES wallboards.

Wallboard No.	Lateral Loading Condition	CL-O (kN)	FL-O (kN)	UL (kN)
1-1	cyclic	54.29	88.48	93.20
1-2	cyclic	39.73	69.14	78.54
1-3	monotonic	36.00	42.00	62.31
1-4	monotonic	18.61	27.67	47.52
2-1	monotonic	-	-	-
2-2	monotonic	41.50	78.81	81.26
2-3	monotonic	20.60	42.00	47.55
2-4	monotonic	21.70	39.00	42.00

Note: CL-O is the load corresponding to the first cracking, as based on observation; FL-O is the load corresponding to the formation of a complete cracking pattern, as based on observation; and UL is the ultimate load at the end of the test.

From the experimental phenomena of the HMES wallboards, the following comment regarding their future design and engineering application can be made: the first crack occurs at the weakest zone in the wallboards where the maximum stress exists. Clearly, the weakest zone is dependent on the configuration of the wallboard. Importantly, the cracking pattern of the wallboard provides a reference for its configurational improvement. For instance, for the wallboard with a door opening, the cracking zone between the top edges of the wallboard and the door indicates that some strengthening measures may be taken there; additionally, some strengthening measures may be taken in the cracking zone between the door and window openings. Therefore, efforts should be made to ensure a shearing working state for the engineering application of this wallboard.

4. Data Analysis

4.1. FE Model

As shown in Figure 9, the HMES wallboards were modeled in ABAQUS/Explicit with C3D8R elements. The C3D8R element is a general-purpose linear brick element, with reduced integration (1 integration point) [28]. The shape functions are the same as for the C3D8 element. In the force-control loading pattern, a uniform increasing pressure is imposed on the top face of specimens gradually until failure occurs. The load capacity of two insulation layers is neglected in the FE (Finite element) model.

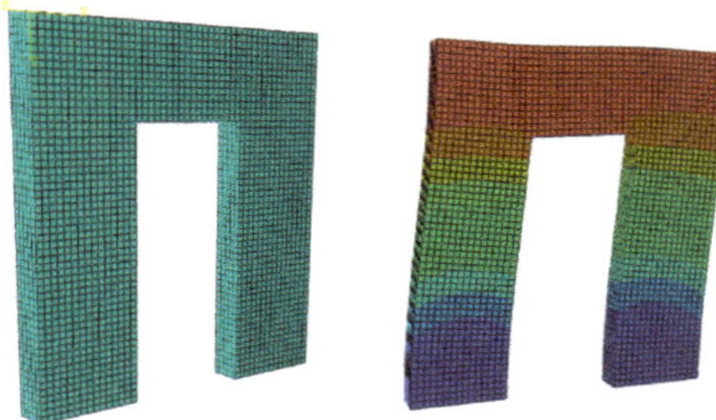

Figure 9. FE model and deformation pattern of specimen 1-3.

The elastic modulus of husk mortar was set as 4976 MPa according to the test results mentioned above. A preliminary FE model with elastic modulus of husk mortar was established to validate the results of HMES wallboards.

Figure 10 shows that there is a relatively reasonable agreement between numerical and test values at the initial stage of the load displacement curve. Such results verify the accuracy of test methods and test results in the elastic range. However, the numerical and test results do not match well after the displacement reaches 6.41 mm. When the specimen 1-3 enters the plastic stage, the elastic constitutive model of husk mortar cannot reflect the plastic behavior of HMES wallboards. A reasonable constitutive model should be adopted for plastic analysis. This issue requires additional research.

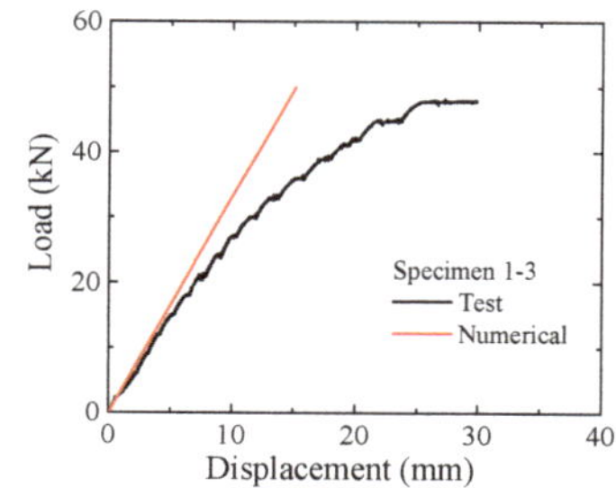

Figure 10. Numerical and test results of specimen 1-3.

4.2. The Characteristic Parameters

Typically, the first cracks are a local failure of the wallboard. With cracking propagation, local failure becomes structural failure. This development might raise the question of how to define structural failure based on the propagation of local failure/cracking, that is, how to delineate the boundary between structural and local failures. In other words, the stressing state of a structure should be expressed by the response data of key points from the experimental measurement or numerical simulation. This issue leads to an updated definition of structural failure. In this definition, the structural failure state is defined as the working state after the structure loses its stable/normal working state, which is not a structural collapse corresponding to the ultimate load. Here, the wallboards are analyzed to reveal their unseen working characteristics and to define their failure loads based on an empirical modelling of the experimental data.

First, a characteristic parameter $E_{j,norm}$ is defined as the normalized approximate strain energy (NASE) sum of the wallboard to the jth load level F_j; that is,

$$E_{j,norm} = \frac{\sum_{i=1}^{N} E_{ij}}{E_M}, \qquad E_{ij} = \frac{1}{2}\overline{E}\varepsilon_{ij}^2 \tag{1}$$

where E_{ij} is the ASE value of the ith measuring point to the jth load increment; N is the number of measuring points distributed on the wallboard; E_M is the maximum sum of strain energy among the measuring points to the individual load increments or the ASE sum to the last (Mth) load increment; $\overline{E}$ is the elastic modulus; and ε_{ij} is the strain value at the ith point to the jth load increment.

For Equation (1), two points must be indicated:

(a) E_{ij} is just an approximate strain energy parameter of the wallboard since it is calculated using a limited number of experimental strain values. Furthermore, $E_{j,norm}$ approximately reflects the global behavior of the wallboard under lateral load. However, $E_{j,norm}$ might better reflect the real working behavior of the wallboard than conventional numerical simulation because of the direct application of experimental data.

(b) At present, a large amount of experimental strain data has not been fully used in the field of structural analysis. Hence, the introduction of $E_{j,norm}$ represents a methodological attempt to reflect some unseen global working behavior of the wallboard (it could be as a structure) by applying experimental strain values.

4.3. Characteristic Curve

For the seven specimens for which the uniformly distributed strain values are measured, Figure 11a–g draws their characteristic curves of the relationship between F_j and $E_{j,norm}$. To be specific, CP stands for the crack point, and FP stands for the failure point, which is defined as the point where the slope of the curve changes greatly.

Figure 11 shows that the seven specimens basically embody a similar feature of three working stages, particularly the failure loads dotted in the $F_j - E_{j,norm}$ curves. Three working stages can be seen in Figure 11, together with the experimental observations:

In Figure 11a, before the load reaches 55.0 kN, specimen 1-1 works in an elastic state. Afterwards, specimen 1-1 enters an elastic-plastic working state, as the inflection point from a linear shape to a nonlinear shape in the $F_j - E_{j,norm}$ curve emerges. As the load changes from 55.0 kN to 65.0 kN, specimen 1-1 works in an elastic-plastic state due to the propagation of the first crack and the appearance of other cracks. Notably, however, specimen 1-1 maintains a stable mode of strain energy distribution in these two working states. When the load reaches 65.0 kN, the second inflection point in the $F_j - E_{j,norm}$ curve emerges, at which point specimen 1-1 cannot maintain a stable mode of strain energy distribution before collapsing. According to the definition in reference [29,30], defining the first inflection point as the cracking load, and the second inflection point as the failure load, it can be seen from Figure 11a that 55.0 kN is the crack load of specimen 1-1, 65.0 kN is the failure load of specimen 1-1, and 93.2 kN is the ultimate load of specimen 1-1.

Similarly, in Figure 11b the crack load, failure load, and ultimate load of specimen 1-2 are 20.5 kN, 69.8 kN, and 78.6 kN, respectively. In Figure 11c the crack load, failure load, and ultimate load of specimen 1-3 are 17.9 kN, 42.0 kN, and 62.3 kN, respectively. In Figure 11d the crack load, failure load, and ultimate load of specimen 1-4 are 18.1 kN, 45.4 kN, and 47.5 kN, respectively. In Figure 11e the crack load, failure load, and ultimate load of specimen 2-2 are 21.5 kN, 74.6 kN, and 81.3 kN, respectively. In Figure 11f the crack load, failure load, and ultimate load of specimen 2-3 are 21.3 kN, 42.5 kN, and 47.6 kN, respectively. In Figure 11g the crack load, failure load, and ultimate load of specimen 2-4 are 27.5 kN, 36.6 kN, and 42.0 kN, respectively.

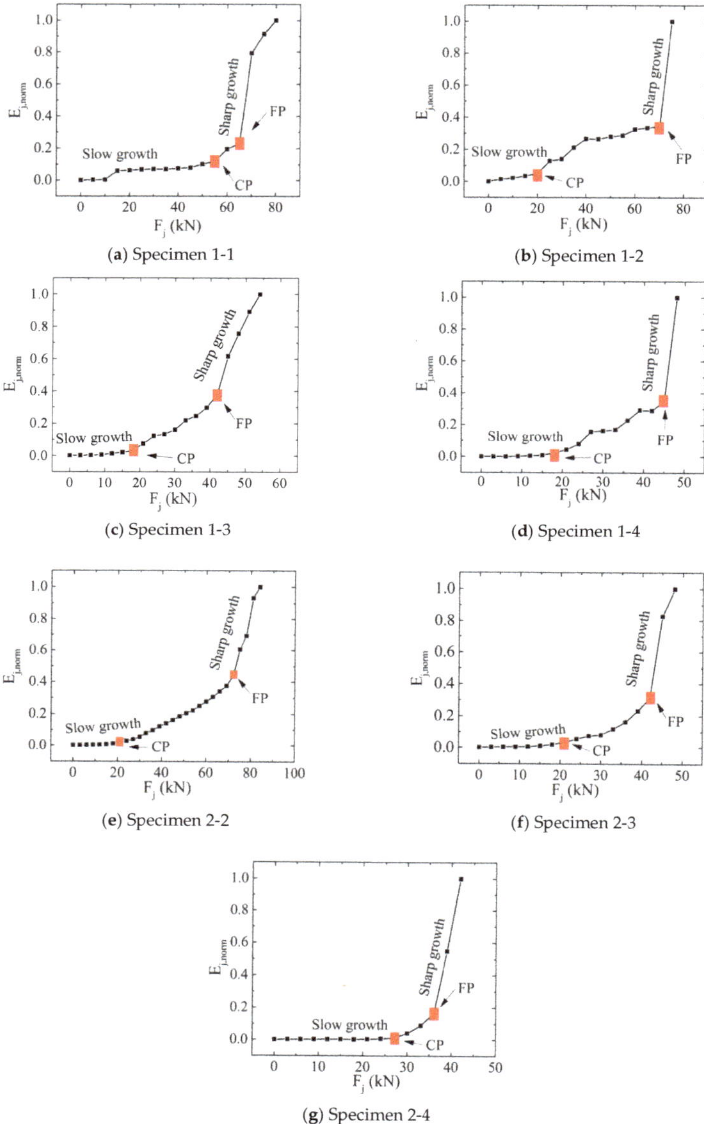

Figure 11. The $F_j - E_{j,\mathrm{norm}}$ curves of the specimens. Note: CP is the cracking load point; FP is the failure load point.

Here, three points must be emphasized: (1) FP exists in the $F_j - E_{j,\mathrm{norm}}$ curve of every specimen, as shown in Figure 11. (2) FP emerges in the elastic-plastic working stage of the specimen. (3) The working states of the wallboard are different before and after the appearance of FP. The former state is relatively stable and maintains a relatively constant mode/pattern, while the latter state is unstable/changeable with the load increment. Therefore, based on the characteristic point in the $F_j - E_{j,\text{norm}}$ curve, the response of a wallboard to a load increment can be divided into two working states: the stable/normal working state occurring until the characteristic point and the unstable/failure working state from the characteristic point to the collapse of the wallboard. The failure load is defined as the load value at the characteristic point.

The results of the crack load and failure load derived from the strain data are consistent with the results based on observations, as shown in Figure 12. Actually, the crack results based on observation requires manpower and depends on the experimental environment. Therefore, the crack load cannot be defined clearly in this way. Moreover, when the crack occurs in the initial stage, the width of the crack is not large enough for observation. With strain analysis method, the initial stage of crack could be captured reasonably.

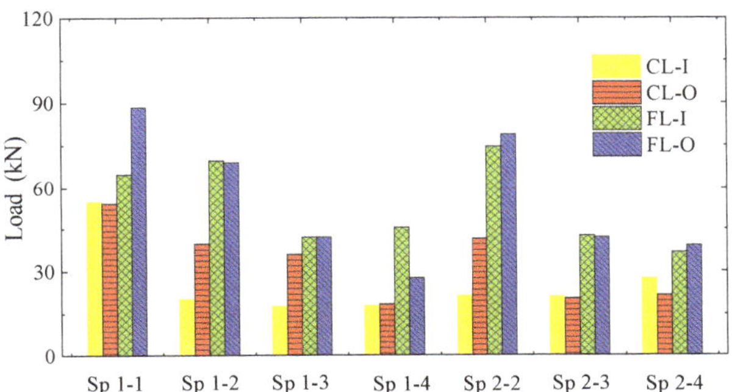

Figure 12. Crack load and failure load with different methods. Note: CL-O is the load corresponding to the first cracking, as based on observation; FL- O is the load corresponding to the formation of a complete cracking pattern, as based on observation; CL-I is the cracking load based on the strain analysis; and FL-I is the failure load based on the strain analysis.

The analysis of the strain data provides a proper way to judge whether cracks occur/propagate in the specimen. The appearance of CP marks that cracks have occurred and have begun to propagate in the specimen.

4.4. MK Criterion

To distinguish the stressing state jump of the structure via the $F_j - E_{j,\text{norm}}$ curve, the MK (Mann-Kenddall) method is applied [24]. The function of F_j is monotonically increasing because the loading scheme is force controlled. The characteristic point defined by the MK method can be used to describe the structural stressing state.

The sequence of E_i is statistically independent. Then, a statistical quantity d_k at the kth load can be defined as

$$d_k = \sum_i^k m_i \ (2 \leq k \leq n) \quad m_i = \begin{cases} +1 & E(i) > E(j) \ (1 \leq j \leq i) \\ 0 & \text{otherwise} \end{cases} \tag{2}$$

where m_i is the cumulative number of the samples and $+1$ indicates adding one to the existing value if the inequality on the right side is satisfied for the jth comparison. Next, calculate the mean value $E(d_k)$ and $Var(d_k)$ of d_k:

$$E(d_k) = k(k-1)/4 \ (2 \leq k \leq n)$$
$$Var(d_k) = k(k-1)(2k+5)/72 \ (2 \leq k \leq n) \tag{3}$$

A new statistical quantity UF_k is defined by Equation (4), and the $UF_k - F$ curve can be plotted.

$$UF_k = \begin{cases} 0 & k = 1 \\ d_k - E(d_k)\ \sqrt{Var(d_k)} & 2 \leq k \leq n \end{cases} \tag{4}$$

For the inverse sequence of E_i, the same steps are conducted to yield the $UB_k - F$ curve. The intersection point of the $UF_k - F$ and $UB_k - F$ curves is the characteristic point of the structural stressing state. In this way, the failure point can be defined properly.

For example, the characteristic point of specimen 1-3 was analyzed, as shown in Figure 13.

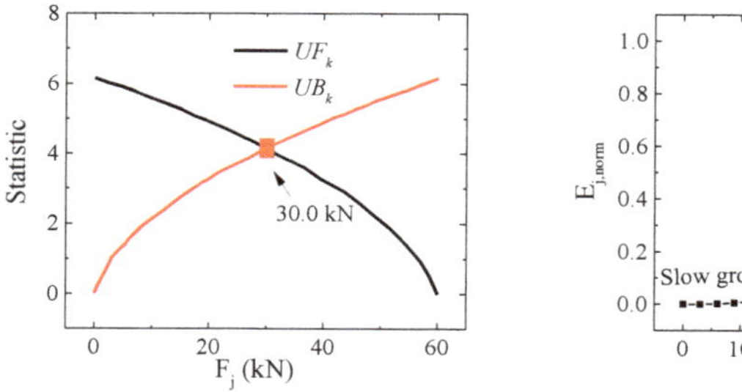

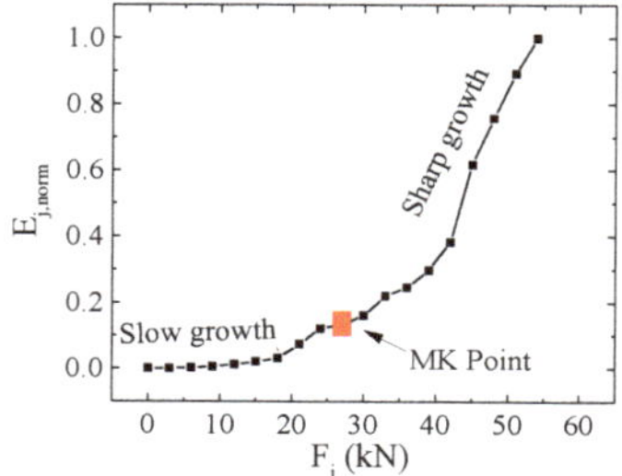

Figure 13. Crack load defined by the MK method.

The other six specimens were correspondingly analyzed with the MK method, and the loads at the MK point are recorded in Figure 14. When CL-O is considered as the standard value, then the load value derived from the MK method is closer than CL-I, as seen from Figure 14.

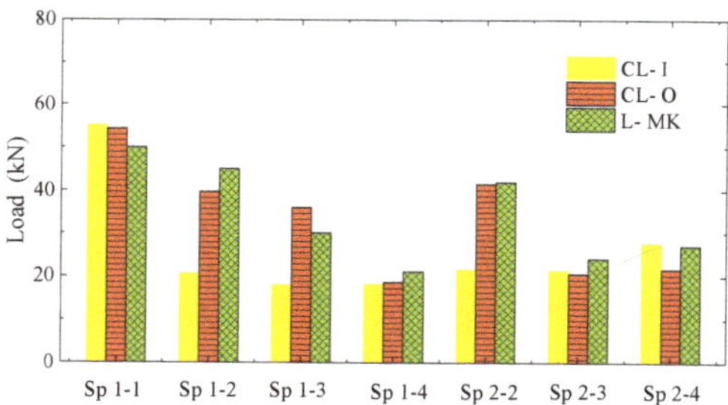

Figure 14. Loads defined by the MK method.

4.5. Structural Stress State

The sum of the strain energies can reflect the overall working state of the specimens. Moreover, the change in the stress state, which reflects the strain distribution, can be derived from the strain data as well. To be specific, Figure 15 shows that strain distribution of specimen 1-3.

Figure 15 indicates that the jump features of the stressing state modes are consistent with those revealed in Figure 11. Figure 15 shows sudden and sharp increases at 18.0 kN and 45.0 kN, with the latter more apparent. Thus, it can be inferred that when the load reaches 21.0 kN and 48.0 kN, the stressing state of the specimen changes to another stressing state sub-mode. Of further note is that the stressing state remains stable between the first characteristic point (P1) and the second characteristic point (P2), whereas the stressing state cannot maintain stability after the second characteristic point.

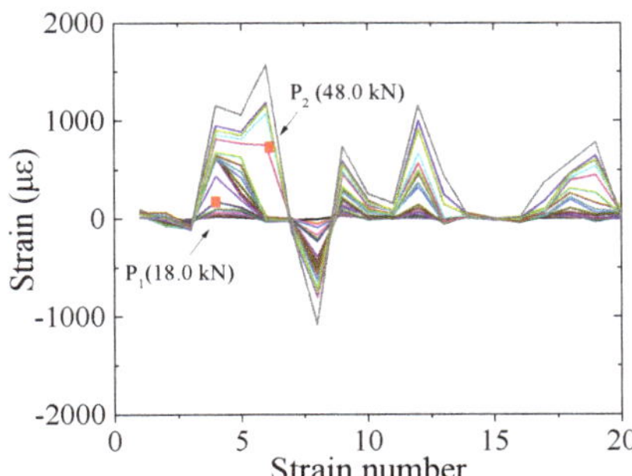

Figure 15. The changing features of strain-based stressing state modes for specimen 1-3.

5. Summary and Conclusions

This study attempts an innovative application of experimental strain data to investigate the working states of HMES wallboards with openings based on the concept of the structural stress state. It should be noted that such investigation is suitable for in-plane shear behavior of infill walls subjected to the monotonic and cyclic loading. Issues with more complex loads, different stiffnesses of joints, rigidity of joining elements, and various configurations require additional research. The investigation draws the following conclusions:

The experimental strains can be transferred into the approximate strain energy values to build the parameter (the NASE sum) characterizing the HMES wallboard's working state. The evolution of the NASE sum will present the characteristic points of the HMES wallboard's working state, i.e., the crack load and failure load. Then, the MK method is applied to detect the characteristic load from $E_{j,\text{norm}} - F_j$ curve and finds out the crack load (L-MK). The comparison of the characteristic loads verify the rationality of the crack loads (L-MK). The modeling method of experimental strains and the MK method could derive the rational and accurate crack loads and provide the rational reference to update the design code of HMES wallboards. Moreover, it provides the method to analyze the local and whole working behaviors of HMES wallboards.

Author Contributions: Conceptualization, S.S.; Data curation, X.C.; Investigation, X.C.; Methodology, X.C.; Software, X.C.; Supervision, S.S.; Validation, G.Z.; Writing—review & editing, S.S. All authors have read and agreed to the published version of the manuscript.

Funding: This research received no external funding.

Conflicts of Interest: The authors declare no conflict of interest.

References

1. Leslie, P.; Pearce, J.M.; Harrap, R.; Daniel, S. The application of smartphone technology to economic and environmental analysis of building energy conservation strategies. *Int. J. Sustain. Energy* **2012**, *31*, 295–311. [CrossRef]
2. Chen, R. *Research and Application of Rice-Hull Mortar Lightweight Composite Wall Panel*; Harbin Institute of Technology: Harbin, China, 2010.
3. Cai, X.S. *In-Plane Mechanical Properties of New Energy-Saving Infill Walls*; Harbin Institute of Technology: Harbin, China, 2014.
4. Marques, R.; Lourenço P, B. Structural behaviour and design rules of confined masonry walls: Review and proposals. *Constr. Build. Mater.* **2019**, *217*, 137–155. [CrossRef]
5. Lourenco, P.B.; Milani, G.; Tralli, A.; Zucchini, A. Analysis of masonry structures: Review of and recent trends in homogenization techniques. *Can. J. Civ. Eng.* **2007**, *34*, 1443–1457. [CrossRef]
6. Ju, R.S.; Lee, H.J.; Chen, C.C.; Tao, C.C. Experimental Study on Separating Reinforced Concrete Infill Walls from Steel Moment Frames. *Constr. Steel Res.* **2012**, *71*, 119–128. [CrossRef]

7. Gu, X.; Wang, L.; Zhang, W.; Cui, W. Cyclic behaviour of hinged steel frames enhanced by masonry columns and/or infill walls with/without CFRP. *Struct. Infrastruct. Eng.* **2018**, *14*, 1–16. [CrossRef]

8. Jurko, Z.; Vladimir, S.; Ivica, G. Cyclic Testing of a Single Bay Reinforced Concrete Frames with Various Types of Masonry Infill. *Earthq. Eng.* **2013**, *41*, 1131–1149.

9. Peng, B.; Wei, S.; Long, L.; Zheng, Q.; Ma, Y.; Chen, L. Experimental Investigation on the Performance of Historical Squat Masonry Walls Strengthened by UHPC and Reinforced Polymer Mortar Layers. *Appl. Sci.* **2019**, *9*, 2096. [CrossRef]

10. Dong, K.; Sui, Z.-A.; Jiang, J.; Zhou, X. Experimental Study on Seismic Behavior of Masonry Walls Strengthened by Reinforced Mortar Cross Strips. *Sustainability* **2019**, *11*, 4866. [CrossRef]

11. Chi, B.; Yang, X.; Wang, F.; Zhang, Z.; Quan, Y. Experimental Investigation into the Seismic Performance of Fully Grouted Concrete Masonry Walls Using New Prestressing Technology. *Appl. Sci.* **2019**, *9*, 4354. [CrossRef]

12. Hashemi, A.; Mosalam, K.M. Shake-table Experiment on Reinforced Concrete Structure Containing Masonry Infill Wall. *Earthquake Eng. Struct. Dyn.* **2006**, *35*, 1827–1852. [CrossRef]

13. Torelli, G.; D'Ayala, D.; Betti, M.; Bartoli, G. Analytical and numerical seismic assessment of heritage masonry towers. *Bull. Earthq. Eng.* **2019**. [CrossRef]

14. Orlando, M.; Betti, M.; Spinelli, P. Assessment of structural behaviour and seismic retrofitting for an Italian monumental masonry building. *J. Build. Eng.* **2019**, *29*, 101115.

15. Betti, M.; Vignoli, A. Modelling and analysis of a Romanesque church under earthquake loading: Assessment of seismic resistance. *Steel Constr.* **2008**, *30*, 352–367. [CrossRef]

16. Betti, M.; Orlando, M.; Vignoli, A. Static behaviour of an Italian Medieval Castle: Damage assessment by numerical modelling. *Comput. Struct.* **2011**, *89*, 1956–1970. [CrossRef]

17. Betti, M.; Galano, L.; Vignoli, A. Finite Element Modelling for Seismic Assessment of Historic Masonry Buildings. In *Earthquakes and Their Impact on Society*; Springer: Cham, Germany, 2016; pp. 377–415.

18. Bartoli, G.; Betti, M.; Vignoli, A. A numerical study on seismic risk assessment of historic masonry towers: A case study in San Gimignano. *Bull. Earthq. Eng.* **2016**, *14*, 1475–1518. [CrossRef]

19. Ferrero, C.; Lourenço, P.B.; Calderini, C. Nonlinear modeling of unreinforced masonry structures under seismic actions: Validation using a building hit by the 2016 Central Italy earthquake. *Frattura ed Integrità Strutturale* **2020**, *14*, 92–114. [CrossRef]

20. Zhou, G.; Pan, D.; Xu, X.; Rafiq, Y.M. An innovative technique for predicting failure/cracking load of masonry wall panel under lateral load. *J. Comput. Civ. Eng.* **2010**, *24*, 377–387. [CrossRef]

21. Zhang, Y.; Zhou, G.C.; Xiong, Y.; Rafiq, M.Y. Techniques for predicting cracking pattern of masonry wallet using artificial neural networks and cellular automata. *J. Comput. Civ. Eng.* **2010**, *24*, 161–172. [CrossRef]

22. Huang, Y.; Zhou, G.; Cui, L.; Zhang, K.; Zhang, M.; Huang, Q.A. Method for Predicting Failure Load of Masonry Wall Panel Based on Generalized Strain Energy Density. *J. Eng. Mech.* **2014**, *140*, 04014061. [CrossRef]

23. Shi, J.; Shen, J.; Zhou, G.; Qin, F.; Li, P. Stressing state analysis of large curvature continuous prestressed concrete box-girder bridge model. *J. Civ. Eng. Manag.* **2019**, *25*, 411–421. [CrossRef]

24. Shi, J.; Li, W.; Zheng, K.; Yang, K.; Zhou, G. Experimental Investigation into stressing state characteristics of large-curvature continuous steel box-girder bridge model. *Constr. Build. Mater.* **2018**, *178*, 574–583. [CrossRef]

25. Shi, J.; Yang, K.; Zheng, K.; Shen, J.; Zhou, G.; Huang, Y. An Investigation into working behavior characteristics of parabolic CFST arches applying structural stressing state theory. *J. Civ. Eng. Manag.* **2019**, *25*, 215–227. [CrossRef]

26. Sun, S.; Ma, D.; Zhou, G. Applications and Analysis of the Composite wall on Construction in Heilongjiang Province. *Procedia Eng.* **2015**, *118*, 160–168. [CrossRef]

27. Ministry of Construction of China. *Standard of Test Method of Mechanical Properties on Ordinary Concrete (GB/T50081-2002)*; Ministry of Construction of China: Beijing, China, 2002.

28. Abaqus, G. *Abaqus 6.11*; Dassault Systemes Simulia Corporation: Johnston, RI, USA, 2011.

29. Zhang, M.; Zhou, G.; Huang, Y.; Zhi, X.; Zhang, D.Y. An ESED Method for Investigating Seismic Behavior of Single-Layer Spherical Reticulated Shell. *Earthq. Struct.* **2017**, *13*, 455–464.

30. Zhang, M.; Parke, G.; Tian, S.; Huang, Y.; Zhou, G. Criterion for Judging Seismic Failure of Suspen-dome Based on Strain Energy Density. *Earthq. Struct.* **2018**, *15*, 123–132.

© 2020 by the authors. Licensee MDPI, Basel, Switzerland. This article is an open access article distributed under the terms and conditions of the Creative Commons Attribution (CC BY) license (http://creativecommons.org/licenses/by/4.0/).

Article

Numerical Study of Bond Slip between Section Steel and Recycled Aggregate Concrete with Full Replacement Ratio

Chao Liu [1,2,*], Lu Xing [1], Huawei Liu [1], Zonggang Quan [2,3], Guangming Fu [2,3], Jian Wu [4], Zhenyuan Lv [1] and Chao Zhu [1,2]

[1] College of Science, Xi'an University of Architecture and Technology, Xi'an 710055, China; xinglu@live.xauat.edu.cn (L.X.); liuhuawei@xauat.edu.cn (H.L.); lvzhenyuan@live.xauat.edu.cn (Z.L.); zhuchao@xauat.edu.cn (C.Z.)
[2] Xi'an Engineering Technology Research Center, Xi'an 710055, China; quanzonggang@xauat.edu.cn (Z.Q.); fuguangming@xauat.edu.cn (G.F.)
[3] Xi'an Research & Design Institute of Wall and Roof Materials, Xi'an 710061, China
[4] Shaanxi Key Laboratory of Safety and Durability of Concrete Structures, Xijing University, Xi'an 710123, China; wujian@xauat.edu.cn
* Correspondence: chaoliu@xauat.edu.cn; Tel.: +86-180-9256-1062

Received: 31 December 2019; Accepted: 19 January 2020; Published: 29 January 2020

Abstract: In this paper, the bond deterioration mechanism of recycled aggregate concrete (RAC) with a full replacement ratio was studied through experimental and numerical simulations. To study the bond behavior and the bond slip between section steel and RAC, nine push-out specimens were designed using the control variable method. The effects of the concrete strength, the embedded length, the cover thickness, and the lateral stirrup ratio on the bond behavior and the bond slip were investigated in detail. The loading process and failure mode of the specimens were observed, and the test curves of the loading end and free end of the specimens were analyzed. Based on the experiment, the finite element method (FEM) was used to simulate the specimens, and the simulation results were analyzed by comparing the experiment data. The analysis of the results showed that the developed model is capable of representing the characteristic bond strength value between section steel and RAC with sufficient accuracy, and the main differences of bond slip between the simulation and the test results are the slippage at the limit state and the moment at which the free end starts to slip.

Keywords: full replacement ratio; section steel and RAC; bond behavior; SRRC (Steel Reinforced Recycled Concrete); bond strength; bond slip; numerical simulation

---

1. Introduction

With the acceleration of industrialization and urbanization, the construction industry has developed rapidly, and the accompanying construction waste has increased dramatically. Due to the large cost of traditional landfill methods and serious environmental pollution, reusing the construction waste resources is imperative [1,2]. As a product of the recycling construction waste, recycled aggregate concrete (RAC) has its own advantages in terms of economy and protection for the environment. RAC is prepared by mixing recycled aggregate [3–6] with a certain proportion and grading, and partially or completely replacing natural aggregates (mainly coarse aggregates). The recycled aggregates are obtained by crushing, cleaning, and classifying the waste concrete.

In recent years, various scholars have conducted investigations on the properties of RAC [7–13]. It was found that RAC with an optimized mix ratio has better mechanical properties compared with ordinary concrete [14,15]. RAC has low weight, which is beneficial for reducing the weight of structure and improving the seismic performance of RAC members. Therefore, the combined application of

RAC and section steel in construction engineering not only reduces the overall weight of a structure, but also solves the problem of poor bearing capacity and rigidity of the structure.

Previous investigations [16–23] indicated that the bond slip behavior between section steel and RAC is an important factor affecting the structural performance, and the numerical simulation method of bond slip is an important theoretical basis of engineering extension and structural calculation analysis. Chen et al. [24] carried out push-out tests on twenty-two specimens of RAC with different replacement rates, and analyzed the effects of the replacement rate, the cover thickness, the bonding site, the hooping ratio, and the particle diameter of the aggregate on the bond strength. However, a mathematical model of the constitutive relationship for bond slip was not proposed. Liu et al. [25] studied the bond slip performance of section steel and RAC under different replacement rates and carried out a push-out test on thirty-six specimens. The constitutive relationship of bond strength under the position function was obtained, and the relationship between replacement rate and average characteristic bond strength was also acquired. Hwang et al. [26] established a numerical model for the bond slip analysis of concrete-filled steel tubular columns, considering the bond slip effect at two nodes. Al-Rousan et al. [27] established a new bond slip model for fiber-reinforced concrete, characterized by the fact that the model was based on anchored carbon fiber rebar and fiber-reinforced concrete.

In general, the research studies on the bond slip and numerical simulation between section steel and RAC are incomplete, especially for section steel and RAC with a full replacement ratio. Further study is important for determining the theoretical and engineering relationships between section steel and RAC.

In this study, the impact factors of bond slip strength between section steel and RAC with a full replacement ratio and the constitutive relationship of bond slip were studied. Moreover, reliable numerical simulation was provided to solve engineering challenges and to be used by other researchers and engineers.

2. Experimental Programs

2.1. Materials

In the study, the RAC with full replacement ratio means that all of the natural coarse aggregates in the concrete have been replaced by recycled coarse aggregates. The specimens were obtained from PC32.5R Portland cement, recycled coarse aggregate, natural fine aggregate, natural fine sand, and urban tap water. The particle size of the recycled coarse aggregate was ranged 5–31.5 mm. The physical properties of the recycled coarse aggregate [28,29] were measured, and the results are shown in Table 1. The obtained results satisfied the requirements of "Recycled Coarse Aggregate for Concrete" GB/T 25177-2010.

Table 1. Physical properties of recycled coarse aggregate.

Bulk Density (kg/m^3)		Needle-Flaky Particle Content (%)	Crushing Indicator (%)	Apparent Density (kg/m^3)	Water Absorption Rate (%)
Close	Loose				
1430	1309	3.90%	17	2458	3.83

Double 10 channel steel plates and double 6mm thick steel plates were bonded with epoxy resin in the experiment. The longitudinally stressed steel provided B16 reinforcement, the stirrup provided A6 reinforcement, and their mechanical performances are shown in Table 2.

Table 2. Mechanical performances of steel and stirrup.

Steel		E_1/MPa	E_2/MPa	ε_y	ε_s	ε_u	f_y/MPa	f_u/MPa
6 mm steel plate		2.06×10^5	0.975×10^5	1738	20,500	29,500	354	425
10 channel steel	Flange	2.08×10^5	0.985×10^5	1690	21,500	29,000	357	420
	web	2.07×10^5	0.967×10^5	1755	21,500	28,500	348	435
A6 reinforcement		2.03×10^5	0.857×10^5	1510	14,500	22,500	310	350
B16 reinforcement		2.06×10^5	0.995×10^5	1895	16,500	30,500	390	530

Note: E_1 is the elastic modulus of the elastic phase; E_2 is the slope of the hardening section of the steel; ε_y is the yield strain of the steel corresponding to f_y; ε_s is the strain of the steel; ε_u is the peak strain of the steel corresponding to f_u; f_y is the yield strength of the steel; f_u is the ultimate strength of the steel.

2.2. Design of Specimens

Nine push-out specimens were designed in the test to study the bond behavior and the bond slip between section steel and RAC. The effects of the concrete strength, the embedded length, the cover thickness, and the lateral stirrup ratio on the bond behavior and the bond slip between section steel and RAC were investigated in detail. The parameters of push-out specimens are shown in Table 3. All strain gauges were arranged at a certain interval on the flange steel plate and the web channel steel. This arrangement did not affect the bonding effect in the interface between section steel and RAC, and ensured the safety and the accuracy of the strain gauge. The section design of specimens and section steel are shown in Figures 1 and 2, respectively.

Table 3. Parameters of push-out specimens.

No.	Concrete Strength	Embedded Length/mm	Cover Thickness/mm	Lateral Stirrup Ratio/%	Stirrup Configuration	f_{cu}/MPa
SRRC-1	C30	740	55	0.2	A6@140	29.9
SRRC-2	C30	540	55	0.2	A6@140	32.43
SRRC-3	C30	740	40	0.2	A6@160	32.43
SRRC-4	C30	740	70	0.2	A6@120	32.43
SRRC-5	C40	740	55	0.2	A6@140	45.12
SRRC-6	C20	740	55	0.2	A6@140	21.60
SRRC-7	C30	740	55	0.25	A6@110	29.9
SRRC-8	C30	740	55	0.3	A6@95	29.9
SRRC-9	C30	940	55	0.2	A6@140	29.9

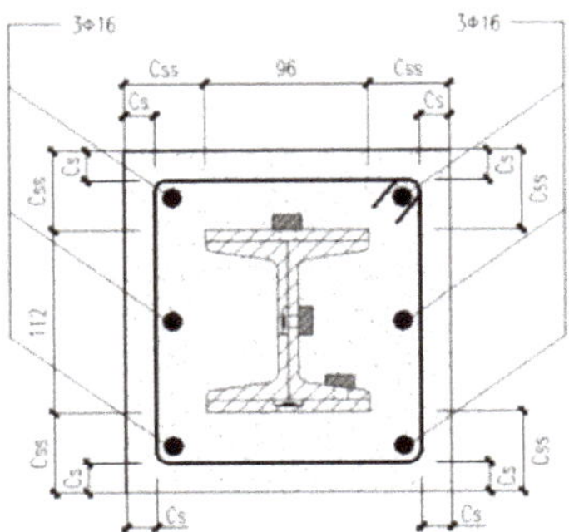

Figure 1. Section design of specimens.

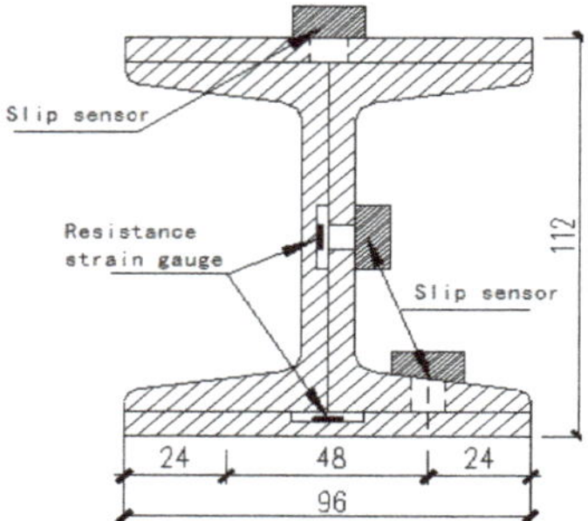

Figure 2. Section design of section steel.

Pre-absorption treatment was conducted on recycled coarse aggregate before preparing RAC [30,31], and the mix design of RAC is shown in Table 4. The reason is that the recycled coarse aggregate is largely porous, which reduces the actual water/cement ratio in the cement slurry and the concrete mix ratio at the same concrete strength.

Table 4. Mix design of recycled aggregate concrete (RAC).

Concrete Strength	Dosage/Kg·m^{-3}				
	Cement	**Recycled Coarse Aggregate**	**Sand**	**Water**	**Additional Water**
C20	335.0	1250.0	650.0	195.0	48.0
C30	370.0	1185.0	660.0	195.0	45.4
C40	425.0	1225.0	640.0	145.0	46.9

Pre-absorption treatment was conducted on recycled coarse aggregate before preparing RAC [30,31]. The reason is that the recycled coarse aggregate with large porosity absorbs a lot of water, and decreases the actual water-cement ratio in the cement slurry. The mix design of RAC is shown in Table 4.

The specimens were made in the seismic engineering laboratory of Xi'an University of Architecture and Technology. The 150 mm × 150 mm × 150 mm cube test blocks were also produced from the same RAC in the test. After the pouring was completed, the push-out specimens were cured in indoor standard conditions (with felt-covered watering and curing). The compressive strength (f_{cu}) of specimens is shown in Table 3.

2.3. Test Method

The strain gauges were applied from dense to sparse distribution along the loading end to the free end, and were bonded to the steel plate by epoxy resin to measure the strain at the flange and the web. Four electronic slip sensors were uniformly arranged on one side of the flange and the web, which were developed by the research team [32], and the Slip-strain (S–ε) relationship of each electronic slip sensor was measured in advance.

The push-out test was carried out on the 2000t compression testing machine in the State Key Laboratory for civil engineering at Xi'an University of Architecture and Technology. Figure 3a,b show photos of the test setup and push-out specimen. The upper end of the specimen was fixed and the lower end was free. A mild steel plate with an "H" hole connected the bottom of the specimen with the loading platform. The topside of the section steel was attached to the compression testing machine with a complete steel plate. A foam pad was laid between the steel support and RAC to ensure flatness, and the loading rate was 0.3 mm/min. Slip occurred initially in the lower part and gradually in the upper part. Therefore, from the perspective of the section steel, the loading end of the specimen was defined at the lower end and the free end was in the upper portion. Double displacement meters were set at the loading end and the free end, respectively, which are shown in Figure 3c.

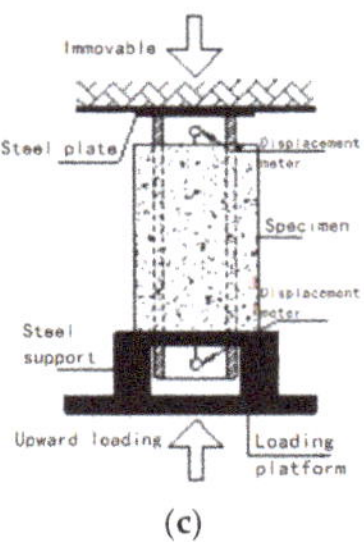

(**a**)   (**b**)   (**c**)

Figure 3. Photos and sketch of the test: (**a**) photo of test setup; (**b**) photo of push-out specimen; (**c**) sketch of push-out specimen.

3. Results and Analysis

3.1. Failure Procedure and Mode

The failure modes of the specimens can be divided into two types: splitting failure mode and bursting failure mode. The failure procedure was roughly as follows: at the initial stage of the specimen loading, there was no obvious change on the surface of each specimen. When the specimen was loaded to 40%–75% of the ultimate load, the initial cracks appeared on the surface of the specimen. With the increase of loading, the initial cracks were mostly concentrated near the loading end at the web, and a small part appeared in the middle. In this time, the initial cracks propagated rapidly, the initial cracks at the loading end extended toward the free end, and the initial cracks in the middle expanded toward both sides as the load increased. When loading to 80%–90% of the ultimate load, the sliding increment of the loading end was obvious and the load increased gently. The initial crack gradually developed into a through crack, and the maximum crack width reached 2–3 mm. As the load continued increasing to the ultimate load, the load sharply dropped to 50%–70% of it, therefore the specimen was considered to be broken by the through crack. If the load continued to increase, the changes were minimal and stabilized with the increasing drifts. It was considered that the load was a residual load. In the process, multiple cracks were generated, in which the original cracks developed secondary cracks, and the damage of RAC increased. The loading ended when the section steel was pushed out 4–6 mm.

The failure mode for SRRC-1, SRRC-4, and SRRC-8 was bursting failure, and the rest specimens showed splitting failure. Splitting failure is the typical failure mode, and here the initial cracks appeared at the loading end of the web sides. With increased load, the cracks extended to the free end and some fine cracks appeared on the specimen gradually. When the load reached the peak load, the initial cracks extended to the upper part of the specimen. Then the load fell rapidly and tended to be gentle gradually. There was a penetrating crack on the flange side and at web sites at this stage, as shown in Figure 4 (taking SRRC-5 as an example).

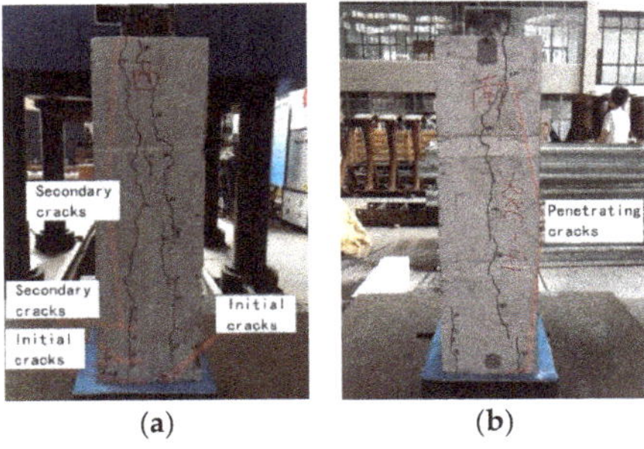

(**a**)   (**b**)

Figure 4. Splitting failure of SRRC-5: (**a**) web; (**b**) flange.

With bursting failure, the initial cracks occurred in the middle of the flange or the web. As the load increased, the initial cracks gradually expanded toward the loading and free ends, and some new fine cracks occurred. When the load reached about 80%–90% of the ultimate load, the initial cracks continued expanding and widening, and there were many obvious secondary cracks. As shown in Figure 5 (taking SRRC-1 as an example), through cracks were present on both the flange and the web sides after failure. This is one of the main features of bursting failure that makes it different from the splitting failure.

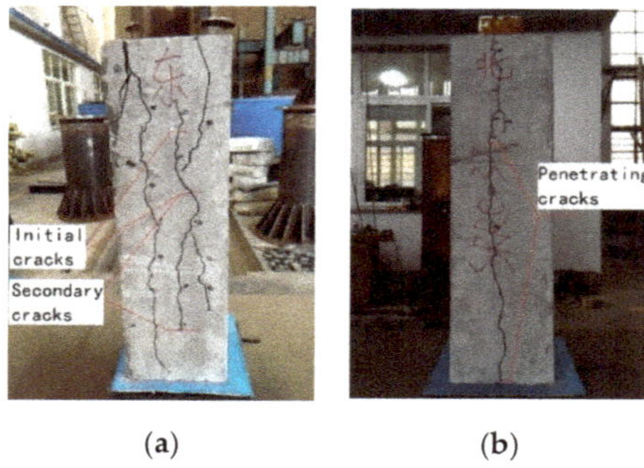

(a) (b)

Figure 5. Bursting failure of SRRC-1: (a) web; (b) flange.

It can be seen from SRRC-4 and SRRC-8 specimens that a high lateral stirrup ratio and high cover thickness make the specimen more prone to bursting failure. The reason for this phenomenon is that a high lateral stirrup ratio and high cover thickness are effective in preventing the deformation of concrete and further improving the cracking load of cracks.

3.2. Characteristic of P–S Curves

The loading end slip curve (P-S curve) can be simplified to the model shown in Figure 6. Here, The load is divided into two categories, each of them showing basically the same changes, which are divided into three parts: rising, sag, and gentle loads. Type (I) is characterized by a large initial load (65%–75% of the peak load), with a residual load that is slightly lower than the initial load. Type (II) is characterized by a small initial load (40%–65% of the peak load), with a residual load that is slightly higher than the initial load. The P–S curves of the specimens are shown in Figure 7.

The following definitions of the characteristic points in Figure 6 are given:

(1) The initial load Ps: The load when obvious slippage occurred on specimens (point A)
(2) The ultimate load Pu: The maximum value of the specimens (point B)
(3) The residual load Pr: The load corresponding to the end of the descending stage (point C)

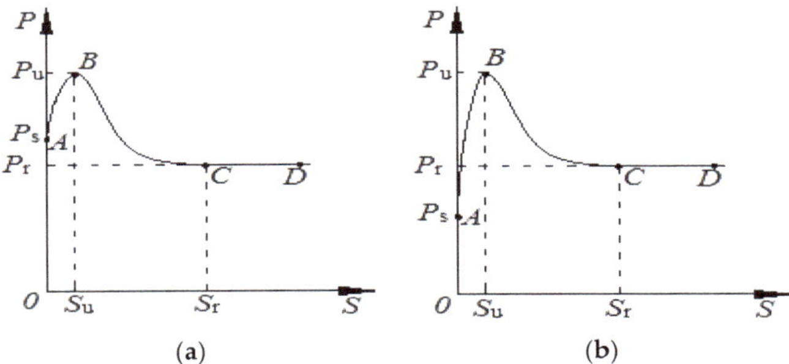

Figure 6. P–S curve models of the loading end: (**a**) Type (I); (**b**) Type (II).

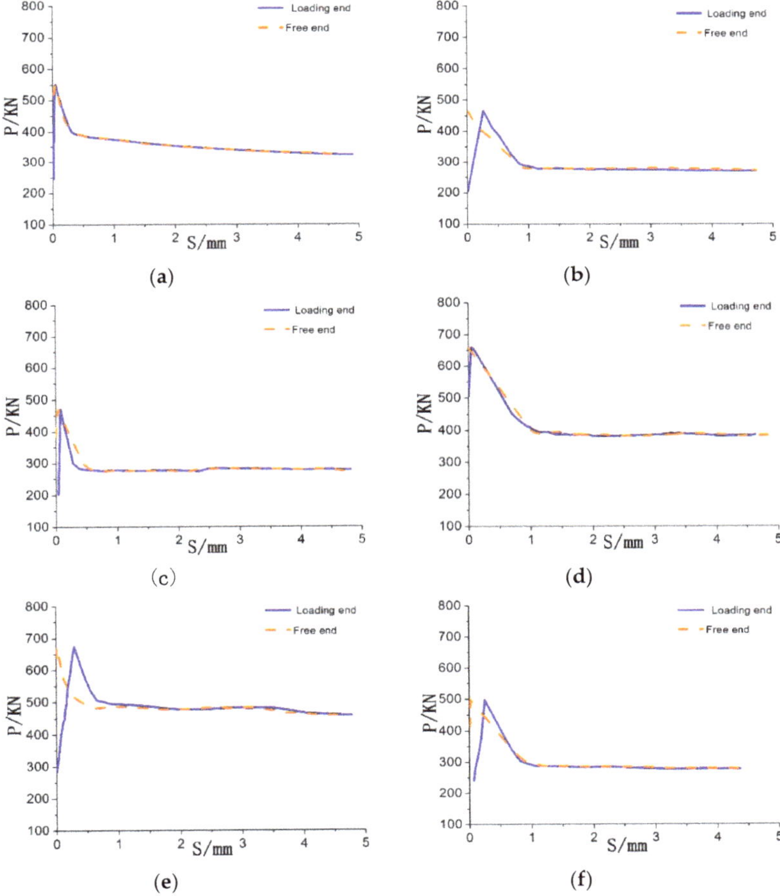

Figure 7. *Cont.*

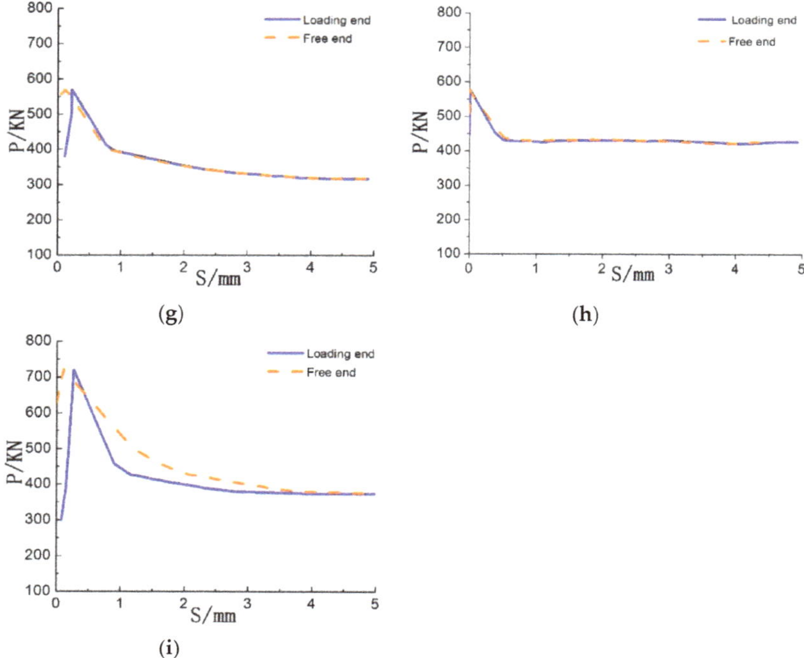

Figure 7. P–S curves of each specimen: (**a**) SRRC-1; (**b**) SRRC-2; (**c**) SRRC-3; (**d**) SRRC-4; (**e**) SRRC-5; (**f**) SRRC-6; (**g**) SRRC-7; (**h**) SRRC-8; (**i**) SRRC-9.

In this paper, the P–S curves of the loading end are divided into four stages: nonslip, slip-crack, descending, and residual.

(1) OA in Figure 6 indicates the nonslip stage of the specimens. The key point is the initial bond load point, which determines the length of the section, and the main load is borne by the chemical adhesive force at this section. The composition of the bond stress in the section steel and RAC is similar to in the section steel and ordinary concrete. The bond stress is caused by chemical adhesion and frictional resistance in the article [33].

(2) AB in Figure 6 indicates the slip-crack stage of the specimens, where the curve is basically a linear relationship and the slope is too large. When loading to 40%–65% of the ultimate load (the load is defined as the initial load P_s, the corresponding bond strength is the average initial bond strength $\bar{\tau}_s$), the loading end of the specimen begins to slip and developed rapidly.

(3) The load drops sharply and the specimen has longitudinal through cracks when loading increases to the ultimate load P_u (the corresponding bond strength is the average limited bond strength $\bar{\tau}_u$). The reason is that the chemical adhesion of the descending stage is suddenly broken and the friction is not sufficient to support the ultimate load.

(4) The residual mainly depends on the residual load. The P–S curve is basically a horizontal line when the load falls to 50%–70% of the ultimate load (the load is defined as the residual load P_r, the corresponding bond strength is the average residual bond strength $\bar{\tau}_r$). It can be concluded that the determinants of each stage are the characteristic loads.

3.3. Influence Analysis of Various Factors

The bond strength between the section steel and RAC can be considered to be evenly distributed along the length of the section steel under the push-out test conditions. The average bond strength can be expressed by Equation (1).

$$\overline{\tau} = \frac{P}{L_e \cdot C} \tag{1}$$

where $\overline{\tau}$ is the average bond stress in MPa; P is the load in N; L_e is the embedded length of section steel in mm; and C is the perimeter of section steel in mm.

3.3.1. Concrete Strength

The bond strength is basically a linear relationship with the tensile strength of RAC, and the bond strength increases with the increase of tensile strength, as shown in Figure 8a. Equation (2) [34] was adopted in this study, which reflects the relationship between the tensile strength and compressive strength of RAC.

$$f_t = 0.18 f_{cu}^{\frac{2}{3}} \tag{2}$$

where f_{cu} is the compressive strength of RAC; f_t is the tensile strength of RAC.

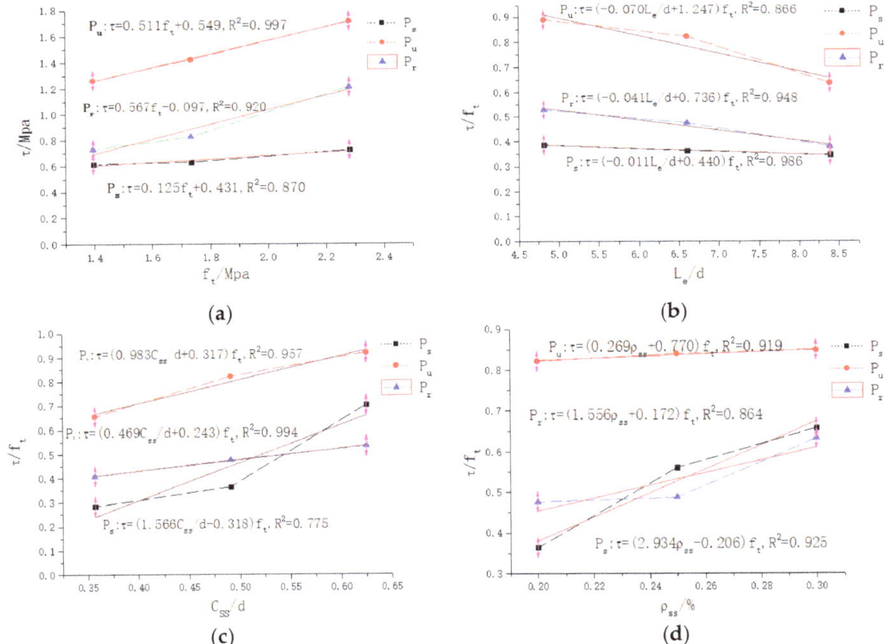

Figure 8. Relationship of bond strength and various factors: (**a**) concrete strength; (**b**) embedded length; (**c**) cover thickness; (**d**) lateral stirrup ratio.

The relationship between the tensile strength (f_t) of RAC and the average bond strength ($\overline{\tau}$) is obtained by statistical regression, which is fit as Equations (3)–(5).

$$\overline{\tau_s} = 0.125 f_t + 0.431 \tag{3}$$

$$\overline{\tau_u} = 0.511 f_t + 0.549 \tag{4}$$

$$\overline{\tau_r} = 0.567 f_t - 0.097 \tag{5}$$

3.3.2. Embedded Length

The relative bond strength is defined as the ratio of the average bond strength to the tensile strength (τ/f_t), and the relative embedded length is defined as the ratio of the embedded length to the height of the section steel (L_e/d). The relationship between the embedded length and the average initial bond strength, the average ultimate bond strength, and the average residual bond strength are shown in the following equations:

$$\overline{\tau_s} = (-0.011 L_e/d + 0.440) f_t \tag{6}$$

$$\overline{\tau_u} = (-0.070 L_e/d + 1.247) f_t \tag{7}$$

$$\overline{\tau_r} = (-0.041 L_e/d + 0.736) f_t \tag{8}$$

As can be seen from Figure 8b, the bond stress decreases as the embedded length increases. The reduction effect of the average initial bond strength is not obvious, and the average ultimate bond strength decreases significantly.

3.3.3. Cover Thickness

The relative cover thickness is calculated from the ratio of the cover thickness to the height of the section steel (C_{ss}/d). The cover thickness refers to the distance between the section steel and the outer surface of RAC. The relationship is shown in the following equations:

$$\overline{\tau_s} = (1.566 C_{ss}/d - 0.318) f_t \tag{9}$$

$$\overline{\tau_u} = (0.983 C_{ss}/d + 0.317) f_t \tag{10}$$

$$\overline{\tau_r} = (0.469 C_{ss}/d + 0.243) f_t \tag{11}$$

It can be seen from Figure 8c that the average characteristic bond strength obviously increases with the increase of the cover thickness.

3.3.4. Lateral Stirrup Ratio

The effect of the lateral stirrup ratio is similar to that of the cover thickness, which can effectively prevent the lateral deformation of the RAC and delay the cracking time. The equations are as follows.

$$\overline{\tau_s} = (2.934 \rho_{ss} - 0.206) f_t \tag{12}$$

$$\overline{\tau_u} = (0.269 \rho_{ss} - 0.770) f_t \tag{13}$$

$$\overline{\tau_r} = (1.556 \rho_{ss} - 0.172) f_t \tag{14}$$

It can be seen from Figure 8d that the average characteristic bond strength increases with the increase of the lateral stirrup ratio, and the average initial bond strength increases obviously. The effect of increasing the ultimate bond strength is poor, indicating that the increase of the lateral stirrup ratio can effectively delay the appearance of the initial crack and increase the cracking load, but the effect on improving the average ultimate bond strength is not significant.

3.4. Formulas

The characteristic bond load and the average characteristic bond strength of specimens are shown in Table 5. The formulas for the average bond stress of the four factors were established by statistical regression analysis. They can be expressed as follows.

$$\bar{\tau}_s = (\frac{-0.686C_{ss}}{d} + \frac{0.020L_e}{d} + 2.506\rho_{sv} + 0.067)f_t \tag{15}$$

$$\bar{\tau}_u = (\frac{-0.335C_{ss}}{d} + \frac{0.015L_e}{d} + 0.718\rho_{sv} + 0.683)f_t \tag{16}$$

$$\bar{\tau}_r = (\frac{-0.493C_{ss}}{d} + \frac{0.006L_e}{d} - 0.842\rho_{sv} + 0.590)f_t \tag{17}$$

where $\bar{\tau}_s$ is average initial bond strength; $\bar{\tau}_u$ is average ultimate bond strength; $\bar{\tau}_r$ is average residual bond strength.

In order to verify the reliability of the formulas, the comparison was performed between the calculation of the formulas and the experiment data from this test, as well as using data from Yin et al. [35], Chen et al. [24], and Chen et al. [36]. The results are shown in Table 6.

Table 6 indicates that the average ultimate bond strength and the average residual bond strength fit well, but the fitting result of the average initial bond strength has a certain error. One of the reasons for this error is the different values of initial load between the man-made and instrument methods. In addition, Equation (2) by Xiao et al. [34] for the tensile strength of RAC was used in this study, but the rest of the articles adopted ordinary concrete formulas. The tensile strength of RAC under the same compressive strength is higher than in this paper.

Table 5. Results of characteristic load and bond stress tests.

| No. | Main Anchoring Condition | | | | | P_s/KN | P_u/KN | P_r/KN | $\bar{\tau}_s$/MPa | $\bar{\tau}_u$/MPa | $\bar{\tau}_r$/MPa |
	Replacement (%)	Concrete Strength (MPa)	Cover Thickness (mm)	Embedded Length (mm)	Lateral Stirrup Ratio (%)						
SRRC-1	100	C30	55	740	0.2	248	559	324	0.633	1.426	0.827
SRRC-2	100	C30	55	540	0.2	203	466	278	0.710	1.629	0.972
SRRC-3	100	C30	40	740	0.2	204	471	275	0.520	1.202	0.702
SRRC-4	100	C30	70	740	0.2	505	660	383	1.288	1.684	0.977
SRRC-5	100	C40	55	740	0.2	284	673	478	0.724	1.717	1.219
SRRC-6	100	C20	55	740	0.2	242	497	287	0.617	1.268	0.732
SRRC-7	100	C30	55	740	0.25	381	571	332	0.972	1.457	0.847
SRRC-8	100	C30	55	740	0.3	447	577	430	1.140	1.472	1.097
SRRC-9	100	C30	55	940	0.2	300	721	376	0.602	1.448	0.755

Table 6. Comparison of characteristic bond strength.

| Source | No. | Initial Bond Strength | | Calculated/Tested | Limit Bond Strength | | Calculated/Tested | Residual Bond Strength | | Calculated/Tested |
		Calculated	Tested		Calculated	Tested		Calculated	Tested	
Article	SRRC-1	0.630	0.633	0.996	1.320	1.426	0.926	0.827	0.827	1.000
	SRRC-2	0.600	0.710	0.845	1.344	1.629	0.825	0.892	0.972	0.918
	SRRC-3	0.833	0.520	1.602	1.475	1.202	1.228	0.993	0.702	1.416
	SRRC-4	0.497	1.288	0.386	1.311	1.684	0.779	0.751	0.977	0.769
	SRRC-5	0.829	0.724	1.144	1.736	1.717	1.011	1.087	1.219	0.892
	SRRC-6	0.507	0.617	0.822	1.063	1.268	0.838	0.665	0.732	0.909
	SRRC-7	0.848	0.972	0.872	1.382	1.457	0.949	0.900	0.847	1.062
	SRRC-8	1.065	1.140	0.934	1.444	1.472	0.981	0.973	1.097	0.887
	SRRC-9	0.731	0.602	1.213	1.442	1.448	0.996	0.853	0.755	1.129
Chen et al.	SRRAC-11	1.438	1.670	0.861	1.719	1.963	0.876	1.258	1.404	0.896
Chen et al.	SRAC-5	1.318	0.930	1.417	1.821	1.510	1.206	1.238	0.990	1.251
Yin et al.	SRRC-34	0.860	0.501	1.717	1.213	0.905	1.340	0.801	0.412	1.944
	SRRC-35	0.807	0.349	2.311	1.125	1.139	0.988	0.798	0.629	1.269
	SRRC-36	0.379	0.492	0.770	1.000	1.038	0.964	0.573	0.637	0.899

4. Numerical Simulation

The simulation of interfacial bond stress is a difficult point in the simulated process of bond slip between section steel and RAC. Nonlinear spring units were utilized to solve the problem in this study, which included two aspects: one was the preparation of the nonlinear syntax for the inp file, and the other was the determination of the constitutive relationship of the spring element.

4.1. Finite Element Model

4.1.1. Element and Material

The solid element C3D8R for section steel and RAC was used in this study, which is an 8-node hexahedron reduction integral element. This element has more accurate results and saves calculation time, and is also suitable for meshing refinement. The linear three-dimensional truss element T3D2 was adopted for steel and stirrups, which has two nodes, each with three degrees of freedom. The steel and stirrups were assigned as truss elements when meshing.

The experimental materials adopted in this study were described as shown in Table 7. The properties of the steel used in the tests were determined in accordance with the "Code for Design of Concrete Structures" [37]. The elastic properties of the second-class coarse aggregate for RAC are shown in Table 8. The plastic damage model of Abaqus was selected for the plastic part of the RAC.

Table 7. Properties of steel.

Material	Elastic Modulus/MPa	Density/Kg/m³	Poisson's Ratio	Plasticity	
				Yield Stress/MPa	Plastic Strain
6 mm Steel plate No. 10 channel steel	206,000	7850	0.3	354	0
A6 Reinforcement (HPB300)	203,000	7850	0.3	310	0
B16 Reinforcement (HRB335)	206,000	7850	0.3	390	0

Table 8. Properties of concrete.

Concrete Strength	Elastic Modulus/MPa	Density/Kg/m³	Poisson's Ratio
C20	18,480	1700	0.2
C30	23,420	1700	0.2
C40	28,510	1700	0.2

4.1.2. Analysis Step and Constraint

The initial incremental step was 0.02, the minimum incremental step was the default 0.00001, and the maximum incremental step was 1000. These were set to meet the calculation requirements in this simulation. In the interaction module, the embedding relationship was defined between the RAC and reinforcement cage, which was made of stirrups and steel. The reference point above the surface of the section steel acting on the displacement load was provided. The loading speed was 0.3 mm/min, and the loading frequency was set as the amplitude. The binding constraints between the bottom slab and the RAC were defined, and the boundary conditions at the bottom of the slab were set as fixed. The specimen adding constraint is shown in Figure 9.

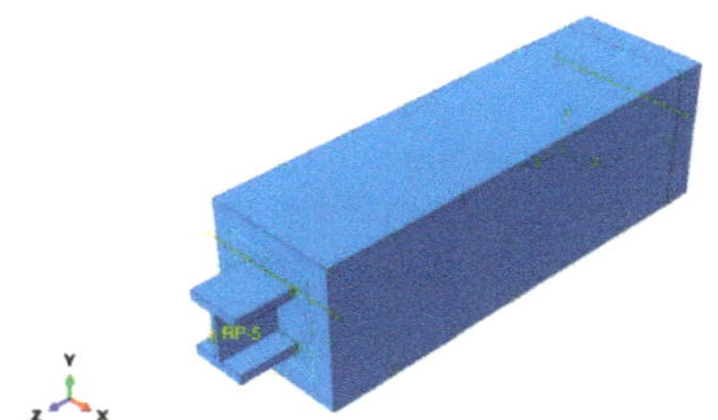

Figure 9. Specimen containing section steel and RAC.

4.1.3. Meshing

The mesh generation of the specimen was performed after the assembly of the components and the constraint settings were completed, which affected the establishment of subsequent nonlinear springs. The intersection interface for the spring element arrangement is most important, and it is located between the section steel and the RAC. The mesh for the interface between the section steel and the RAC was divided in order to successfully arrange the subsequent spring elements. The section steel was cut according to the geometrical axis, as shown in Figure 10, and the section steel and RAC parts were set as independent.

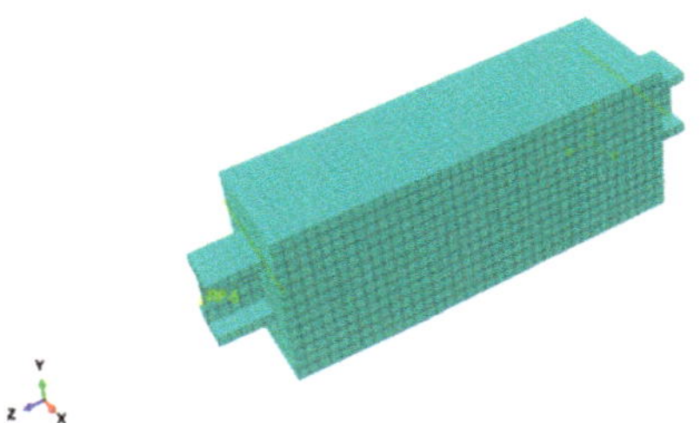

Figure 10. Mesh division for specimen containing section steel and RAC.

The node set of the interface for the section steel and the RAC was established after the mesh was divided, as shown in Figure 11. All nodes in the node set were exported in post-processing and were numbered using the "VLOOKUP" function in excel software.

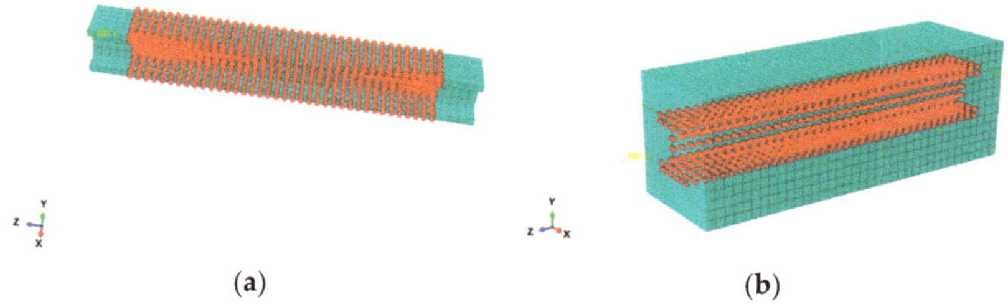

(**a**)   (**b**)

Figure 11. Node set: (**a**) section steel; (**b**) RAC.

Each specimen in this simulation has 1140 nodes. The VLOOKUP function of a corresponding node is

=VLOOKUP(C4&D4&E4,CHOOSE({1,2,3,4},$G$4:$G$1143&$H$4:$H$1143&$I$4:$I$1143,$F$4: $F$1143),2,0).

SRRC-1 is used as an example in Table 9.

Table 9. Relationship of node for SRRC-1.

No.	Section 1	Node 1	Section 2	Node 2	No.	Section 1	Node 1	Section 2	Node 2
1	Tong-1	1	Xinggang-1	15	42	Tong-1	173	Xinggang-1	283
2	Tong-1	2	Xinggang-1	14	43	Tong-1	178	Xinggang-1	1116
3	Tong-1	3	Xinggang-1	13	44	Tong-1	179	Xinggang-1	792
4	Tong-1	4	Xinggang-1	24	45	Tong-1	180	Xinggang-1	791
5	Tong-1	5	Xinggang-1	23	46	Tong-1	181	Xinggang-1	790
6	Tong-1	6	Xinggang-1	22	47	Tong-1	182	Xinggang-1	789
7	Tong-1	7	Xinggang-1	21	48	Tong-1	183	Xinggang-1	1296
8	Tong-1	8	Xinggang-1	20	49	Tong-1	184	Xinggang-1	1224
9	Tong-1	9	Xinggang-1	19	50	Tong-1	185	Xinggang-1	1223
10	Tong-1	10	Xinggang-1	18	51	Tong-1	186	Xinggang-1	1222
11	Tong-1	11	Xinggang-1	17	52	Tong-1	187	Xinggang-1	1260
12	Tong-1	12	Xinggang-1	16	53	Tong-1	188	Xinggang-1	936
13	Tong-1	17	Xinggang-1	499	54	Tong-1	189	Xinggang-1	935
14	Tong-1	18	Xinggang-1	516	55	Tong-1	190	Xinggang-1	934
15	Tong-1	19	Xinggang-1	515	56	Tong-1	191	Xinggang-1	933
16	Tong-1	20	Xinggang-1	514	57	Tong-1	192	Xinggang-1	972
17	Tong-1	21	Xinggang-1	513	58	Tong-1	193	Xinggang-1	1080
18	Tong-1	22	Xinggang-1	512	59	Tong-1	194	Xinggang-1	1079
19	Tong-1	23	Xinggang-1	511	60	Tong-1	195	Xinggang-1	1078
20	Tong-1	24	Xinggang-1	510	61	Tong-1	323	Xinggang-1	320
21	Tong-1	25	Xinggang-1	509	62	Tong-1	324	Xinggang-1	356
22	Tong-1	26	Xinggang-1	508	63	Tong-1	325	Xinggang-1	137
23	Tong-1	27	Xinggang-1	507	64	Tong-1	326	Xinggang-1	68
24	Tong-1	28	Xinggang-1	506	65	Tong-1	327	Xinggang-1	140
25	Tong-1	29	Xinggang-1	505	66	Tong-1	328	Xinggang-1	461
26	Tong-1	30	Xinggang-1	504	67	Tong-1	329	Xinggang-1	392
27	Tong-1	31	Xinggang-1	503	68	Tong-1	330	Xinggang-1	497
28	Tong-1	32	Xinggang-1	502	69	Tong-1	331	Xinggang-1	245
29	Tong-1	33	Xinggang-1	501	70	Tong-1	332	Xinggang-1	176
30	Tong-1	34	Xinggang-1	500	71	Tong-1	333	Xinggang-1	248
31	Tong-1	162	Xinggang-1	319	72	Tong-1	334	Xinggang-1	284
32	Tong-1	163	Xinggang-1	355	73	Tong-1	339	Xinggang-1	1115
33	Tong-1	164	Xinggang-1	138	74	Tong-1	340	Xinggang-1	788
34	Tong-1	165	Xinggang-1	67	75	Tong-1	341	Xinggang-1	787
35	Tong-1	166	Xinggang-1	139	76	Tong-1	342	Xinggang-1	786
36	Tong-1	167	Xinggang-1	462	77	Tong-1	343	Xinggang-1	785
37	Tong-1	168	Xinggang-1	391	78	Tong-1	344	Xinggang-1	1295
38	Tong-1	169	Xinggang-1	498	79	Tong-1	345	Xinggang-1	1221
39	Tong-1	170	Xinggang-1	246	80	Tong-1	346	Xinggang-1	1220
40	Tong-1	171	Xinggang-1	175	81	Tong-1	347	Xinggang-1	1219
41	Tong-1	172	Xinggang-1	247	82	Tong-1	348	Xinggang-1	1259

Note: Table cells are expressed in absolute form; C4, D4, and E4 are the columns where the x, y, and z coordinates of Section 1 are located. G4: G1143 represents the query columns, which correspond to the C4 column of Section 1; the rest can be deduced by analogy. F4: F1143 is the column for the output formula.

4.1.4. Plastic Damage Model

The premise for studying the plastic damage model for RAC is reflected in the constitutive relationship. The constitutive relationship between the compression and tension in RAC is basically the same as that of ordinary concrete. The specific differences are reflected in several coefficients related to the replacement rate. The equation proposed by Xiao et al. [38] was used for calculation in this study. The stress and strain at different strength levels are shown in Table 10.

Table 10. Characteristic parameters of RAC.

Concrete Strength	Compressive Strength f_c/MPa	ε_c	Tensile Strength f_t/Mpa	ε_t	Elastic Modulus/MPa
C20	16.408	0.00124	1.396	0.00008	18,480
C30	24.647	0.00147	1.830	0.00010	23,420
C40	34.291	0.00168	2.281	0.00011	28,510

The plastic damage model of concrete is mainly used to provide a universal analysis model for analyzing the structure of concrete under cyclic and dynamic loads. This model is based on plastic and isotropic failure assumptions, and it can be used in unidirectional loading, cyclic loading, and other functions [39].

The evolution of the yield or failure surface is controlled by $\widetilde{\varepsilon}_c^{in}$ and $\widetilde{\varepsilon}_t^{in}$, where $\widetilde{\varepsilon}_c^{in}$ represents a compressive inelastic strain and $\widetilde{\varepsilon}_t^{in}$ represents a tensile inelastic strain.

$$\widetilde{\varepsilon}_c^{in} = \varepsilon_c - \varepsilon_{0c}^{el} \tag{18}$$

$$\varepsilon_{0c}^{el} = \sigma_c / E_0 \tag{19}$$

Let the plastic strain in the inelastic strain be $\widetilde{\varepsilon}_c^{pl}$, then the proportion of the inelastic strain is β_c, according to the Abaqus user manual:

$$\widetilde{\varepsilon}_c^{pl} = \widetilde{\varepsilon}_c^{in} - \frac{d_c}{1 - d_c} \times \frac{\sigma_c}{E_0} \tag{20}$$

$$d_c = \frac{(1 - \beta_C)\widetilde{\varepsilon}_c^{in} E_0}{\sigma_c + (1 - \beta_c)\widetilde{\varepsilon}_c^{in} E_0} \tag{21}$$

$$d_t = \frac{(1 - \beta_t)\widetilde{\varepsilon}_t^{in} E_0}{\sigma_t + (1 - \beta_t)\widetilde{\varepsilon}_t^{in} E_0} \tag{22}$$

where d_c is the concrete compression damage factor; d_t is the concrete tensile damage factor; σ_c is the peak compressive stress of RAC; σ_t is the peak tensile stress of RAC; β_c is 0.6; β_t is 0.8; $\widetilde{\varepsilon}_c^{in}$ is the inelastic compressive strain of RAC; $\widetilde{\varepsilon}_t^{in}$ is the inelastic tensile strain of RAC.

4.1.5. F–D Relation of Nonlinear Spring Unit

The constitutive relationship of the spring unit was determined before the establishment of the nonlinear spring unit, and was determined according to the constitutive relationship of the average bond strength at the loading end of the specimen. The interface between the section steel and RAC had three directions, namely longitudinal tangential, normal, and transverse tangential directions. The experiments proved that in the case of structural failure: first, the normal and transverse tangential deformation were much smaller than the longitudinal tangential deformation; second, the longitudinal tangential interaction was characterized by the bond slip phenomenon in the section steel and RAC. The Force-Displacement curve (F–D curve) consistent with the longitudinal tangential direction was employed by the spring element constitutive relationship, because the transverse tangential assumption was consistent with the longitudinal tangential interaction. The law-up interaction was set to a spring element with infinite stiffness because it was subjected to pressure and had high stiffness. The spring element F–D relationship is calculated by:

$$F = \tau \times A \tag{23}$$

where A is the area occupied the spring connection surface, with the calculation diagram shown in Figure 12.

The F–D relationship of the springs under each node can be calculated through the bond slip constitutive relation, which was obtained from the test. However, the calculated F–D relationship did not pass through the coordinate origin and did not satisfy the input requirements of Abaqus, so it was processed. It was completely symmetrically processed in its negative direction in order to complete the F–D curve. The F–D relationship is shown in Figure 13 (taking the intermediate node of the spring at SRRC-1 as an example).

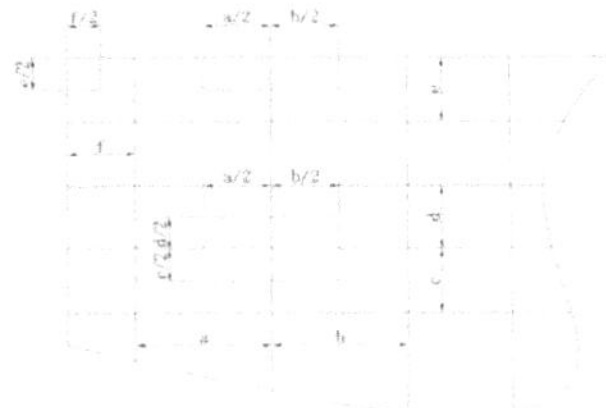

Figure 12. Schematic diagram of the force area calculation for the spring element.

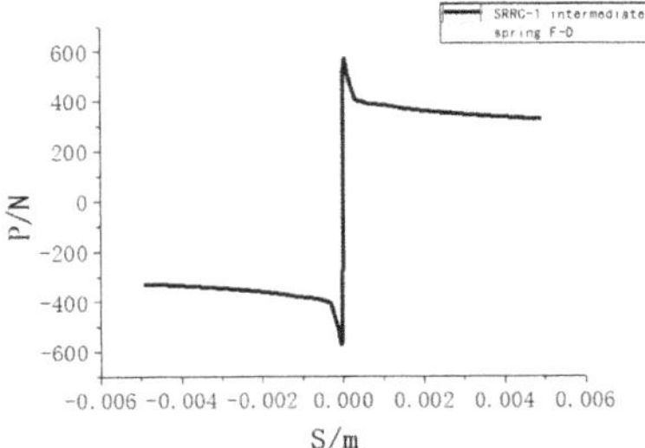

Figure 13. F–D curve for intermediate node spring element of SRRC-1.

4.1.6. Rewriting of Inp File

In this paper, the specific method used in addition of the nonlinear spring unit was as follows: the linear spring was first added to the simulated specimen, then the keywords in the inp file were found, and finally the linear spring was rewritten. The precautions were as follows: first, the rewritten inp file could not be imported into Computer Aided Engineering(CAE) and was applied by Abaqus command operation; second, the added maximum spring force was greater than the maximum force balanced with it; third, the nonlinear stiffness was symmetrically defined in the case of convenience and without affecting the result.

Taking SRRC-1 as an example, the nonlinear spring unit of the rewritten inp file is as follows:

*Spring, elset=Springs/Dashpots-1-spring, nonlinear
1, 1
Nonlinear constitutive relation
*Element, type=Spring2,
elset=Springs/Dashpots-1-spring
Serial number, part one, node one, part two, node two

Nonlinear constitutive relations and corresponding nodes were replaced in order to reduce the length of the article, as shown above. It should be noted that "nonlinear" was required as a keyword after the spring set, indicating that all springs in the set were nonlinear. Two points should be noted when adding nonlinear constitutive relations: one is that the F–D curve needs to pass the coordinate origin, and the other is that the force is connected by "," in the middle of the displacement. For example, in 570, 0.00005, 570 means the force is 570N, and 0.00005 means 0.00005 m. When adding multiple sets of spring sets, the serial number of the corresponding node should be accumulated at once, otherwise the nonlinear stiffness will be overwritten by subsequent coverage, resulting in the failure of multiple sets for spring addition.

In this paper, each set of springs had a total of 1140 nodes and each specimen contained three sets of springs, which were longitudinal tangential, transverse tangential, and normal. Nine sets of specimens were simulated and the results were as outlined in the following subsections.

4.2. Analysis

The model of the nine specimens was simulated. Taking SRRC-1 as an example, the Mises stress nephogram of the section steel and RAC are shown in Figure 14. It can be seen that the normal stress of the section steel gradually decreased from the loading end to the free end, because the stress point was directly coupled with the section steel. The loading end of the RAC is connected to the section steel only through a nonlinear spring, and the free end is directly contact with the steel plate. Therefore, for the RAC, the free end is directly stressed and the normal stress gradually decreases from the free end to the loading end.

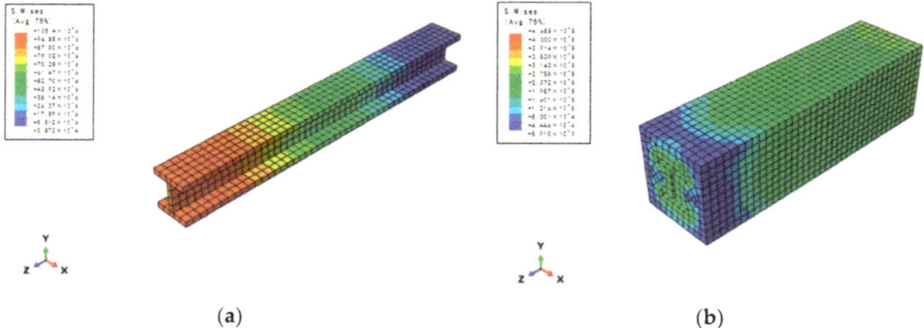

(a)   (b)

Figure 14. Mises stress nephogram of SRRC-1: (**a**) section steel; (**b**) RAC.

This study mainly investigated longitudinal shear stress. The stress nephograms of the specimen in the direction of S23 (S23 stands for shear stress in Abaqus) are shown in Figures 15 and 16. It can be seen that the longitudinal shear stress is relatively uniform distribution along the embedded length. The bond strength is weakest at the outside of the lower flange, and it is greatest at the inside of the lower flange and at the upper flange, which are the main bearing parts of the flange. The bond strength at the inner side of the web is equivalent to the outer side of the lower flange, which is much smaller than the bond strength at the outer side of the web.

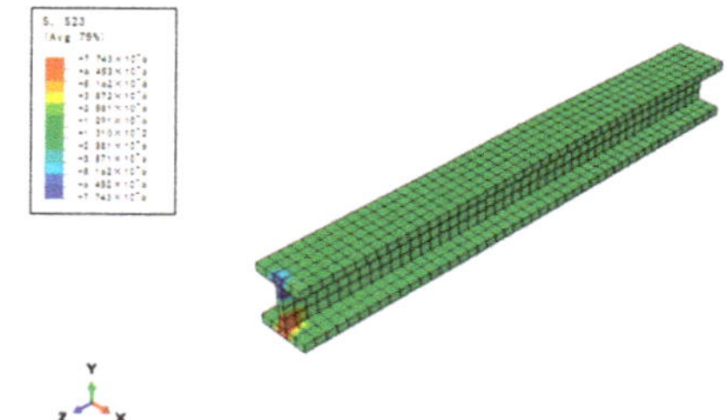

Figure 15. The stress nephograms of section steel in the direction of S23.

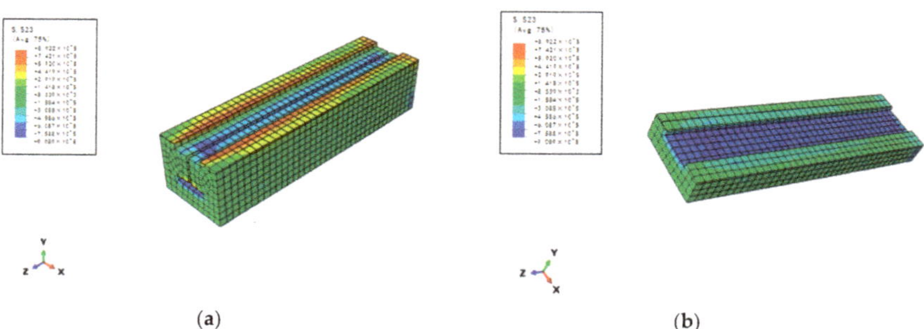

(a)   (b)

Figure 16. *Cont.*

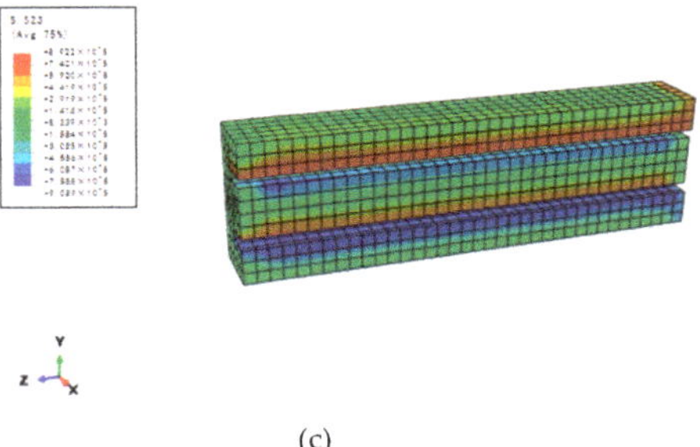

(c)

Figure 16. The cutting stress nephograms of RAC at different directions along the section steel: (**a**) the lower flange; (**b**) the upper flange; (**c**) the web.

According to the stress nephograms, the main failure surfaces in the bond slip of the specimens are the upper flange of the section steel, the inner side of the lower flange, and the outer side of the web. It is added that the bond strength of the outer side of the lower flange and the inner side of the web is much smaller than in the above three sides. Therefore, a certain process must be carried out on the outer side of the section steel flange and the inner side of the web, such as increasing the contact reaction area with the RAC and increasing the mechanical bite force with the RAC. The purpose of this is to enhance the characteristic bond strength of the specimen.

4.2.1. Comparison

As shown in Figure 17, the simulation figure is compared with the experimental figure. It can be seen that the simulation curves are basically consistent with the test curves. The initial load is basically equal with the ultimate load. The residual load has a slight error, which is within ±0.2 MPa and is reasonable. This is consistent with the test showing that SRRC-4, SRRC-7, and SRRC-8 belong to Type (I), with the rest of the specimens belonging to Type (II).

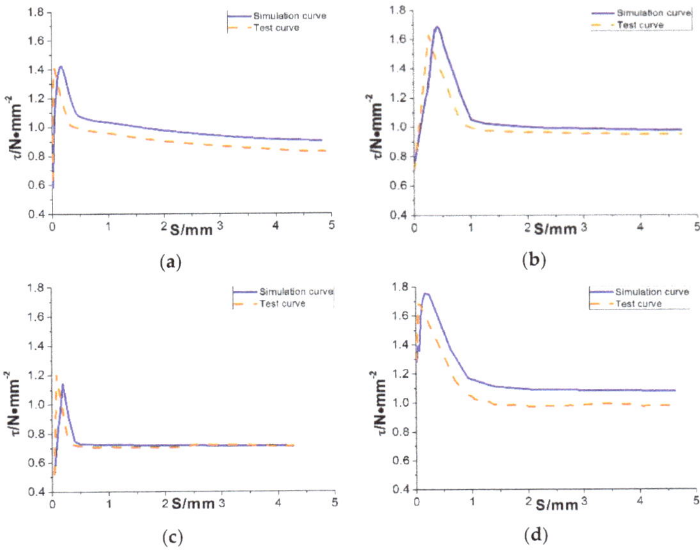

Figure 17. *Cont.*

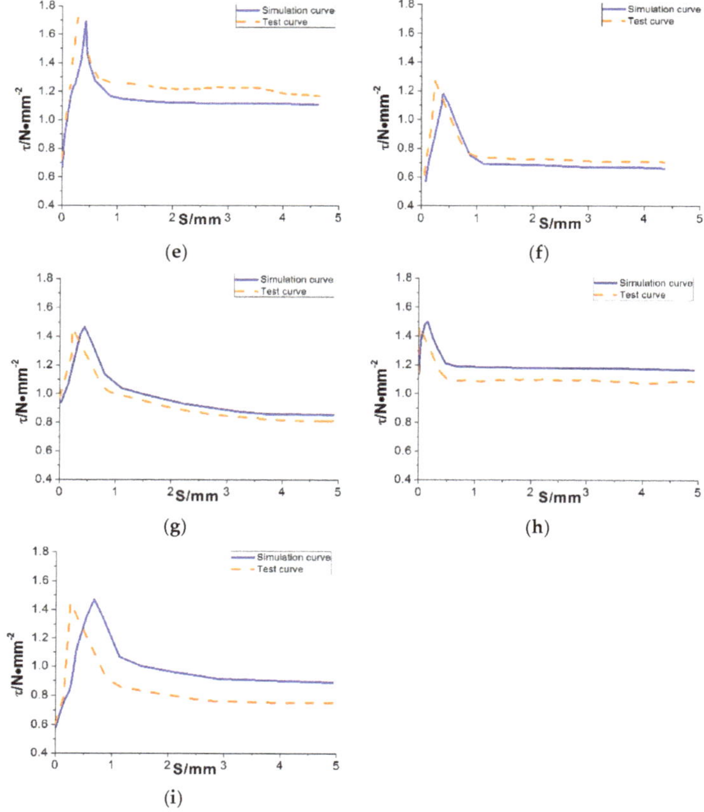

Figure 17. Simulation and test curves of τ-s: (**a**) SRRC-1; (**b**) SRRC-2; (**c**) SRRC-3; (**d**) SRRC-4; (**e**) SRRC-5; (**f**) SRRC-6; (**g**) SRRC-7; (**h**) SRRC-8; (**i**) SRRC-9.

It can be seen that the simulation results of initial bond strength, ultimate bond strength, and residual bond strength are basically equal with the test results. In the limit state, the slip value corresponding to the characteristic bond strength is larger than the test value, and the difference is within 0.15–0.65 mm. The slip value is close to the test value in the residual state, and the difference is within 0–0.15 mm.

In order to further verify the reliability of the simulation results, the experimental data in Chen et al. [24] and Yang et al. [40] were simulated by the numerical simulation method used in this study, and similar results were obtained.

4.2.2. Error Analysis

There are two reasons that the slip value corresponding to the characteristic bond strength is larger than the test value in the limited state:

(1)   The bond stress–slip constitutive relationship between section steel and RAC is related to the embedded length of the section steel [32]. However, the constitutive relationship adopted in this study does not consider the influence of the position function, so there is an error in the slip value.

(2)   From P–S curves of each specimen, it was found that the loading end and the free end slipped almost simultaneously, however, the free end began to slip when the loading end had reached the limited load. The finite element method (FEM) was also a factor that caused error.

5. Conclusions

In this study, nine push-out specimens were designed to study the bond behavior and the bond slip between section steel and RAC, and four factor effects of the concrete strength were investigated. Numerical analysis was conducted, and the simulation and test data were analyzed. The results are summarized as follows:

(1)　The specimens were divided into splitting failure and bursting failure modes. The former is a typical failure mode, where the initial crack starts from the loading end and gradually extends to the free end. The latter is an atypical failure mode, where the initial crack in the middle of the flange side of the specimen gradually extends to both ends.

(2)　The P–S curves were analyzed and classified into Type (I) and Type (II) according to the characteristic load. The initial load of the former is greater than the residual load value, and the latter is smaller than the residual load value. Type (II) occurs more easily due to increases of the cover thickness and the lateral stirrup ratio.

(3)　The relationships between the characteristic bond strength and the concrete strength, the embedded length, the cover thickness, and the lateral stirrup ratio were analyzed. The characteristic bond strength increased with the increase of the concrete strength, the cover thickness, and the lateral stirrup ratio, and it decreased with the increase of the embedded length of the section steel.

(4)　The FEM was used to simulate the specimens, and the simulation results were analyzed by comparing them with the experiment data. The analysis of the results showed that developed model is capable of representing the characteristic bond strength value between section steel and RAC with sufficient accuracy, and the main differences of bond slip between the simulation and the test results are the slippage at the limit state and the moment at which the free end starts to slip.

Author Contributions: Writing—review and editing, C.L., L.X. and H.L.; Resources, Z.Q., G.F., J.W., Z.L. and C.Z. All authors have read and agree to the published version of the manuscript.

Funding: This research was funded by [the National Natural Science Foundation of China] grant number [51878546], [the Innovative Talent Promotion Plan of Shaanxi Province] grant number [2018KJXX-056], [the Key Research and Development Program of Shaanxi Province] grant number [2018ZDCXL-SF-03-03-02], [the Key Laboratory Project of Shaanxi Province] grant number [XJKFJJ201904], [the Science and Technology Innovation Base of Shaanxi Province "Technology Innovation Service Platform for Solid Waste Resources and Energy Saving Wall Materials"] grant number [2017KTPT-19].

Acknowledgments: The authors would like to acknowledge the National Natural Science Foundation of China (Grant No. 51878546), the Innovative Talent Promotion Plan of Shaanxi Province (Grant No. 2018KJXX-056), the Key Research and Development Program of Shaanxi Province (Grant No. 2018ZDCXL-SF-03-03-02), the Key Laboratory Project of Shaanxi Province (Grant No. XJKFJJ201904), and the Science and Technology Innovation Base of Shaanxi Province "Technology Innovation Service Platform for Solid Waste Resources and Energy Saving Wall Materials" (2017KTPT-19) for financial support.

Conflicts of Interest: The authors declare no conflict of interest.

References

1.　Muluken, Y.; Kasun, H.; Shahria Alam, M.; Cigdem, E.; Rehan, S. An overview of construction and demolition waste management in Canada: A lifecycle analysis approach to sustainability. *Clean Technol. Environ. Policy* **2013**, *15*, 81–91.

2.　Zhao, W.; Leeftink, R.B.; Rotter, V.S. Evaluation of the economic feasibility for the recycling of construction and demolition waste in China—The case of Chongqing. *Resour. Conserv. Recycl.* **2009**, *54*, 377–389. [CrossRef]

3.　Medjigbodo, S.; Bendimerad, A.Z.; Roziere, E.; Loukil, A. How do recycled concrete aggregates modify the shrinkage and self-healing properties? *Cem. Concr. Compos.* **2018**, *86*, 72–86. [CrossRef]

4.　Silva, R.V.; de Brito, J.; Dhir, R.K. Properties and composition of recycled aggregates from construction and demolition waste suitable for concrete production. *Constr. Build. Mater.* **2014**, *65*, 201–217. [CrossRef]

5.　Wang, Y.; Zhang, S.H.; Niu, D.T.; Su, L.; Luo, D.M. Strength and chloride ion distribution brought by aggregate of basalt fiber reinforced coral aggregate concrete. *Constr. Build. Mater.* **2020**, *234*, 117390. [CrossRef]

6.   Wang, Y.Y.; Huang, J.J.; Wang, D.J.; Liu, Y.F.; Zhao, Z.J.; Liu, J.P. Experimental study on hygrothermal characteristics of coral sand aggregate concrete and aerated concrete under different humidity and temperature conditions. *Constr. Build. Mater.* **2020**, *230*, 117034. [CrossRef]

7.   Liu, C.; Lv, Z.Y.; Zhu, C.; Bai, G.L.; Zhang, Y. Study on Calculation Method of Long Term Deformation of RAC Beam based on Creep Adjustment Coefficient. *KSCE J. Civ. Eng.* **2018**, *23*, 260–267. [CrossRef]

8.   Xiao, J.Z.; Li, W.G.; Fan, Y.H.; Huang, X. An overview of study on recycled aggregate concrete in China (1996–2011). *Constr. Build. Mater.* **2012**, *31*, 364–383. [CrossRef]

9.   Thomas, C.; Setién, J.; Polanco, J.A.; Alaejos, P.; de Juan, M.S. Durability of recycled aggregate concrete. *Constr. Build. Mater.* **2013**, *40*, 1054–1065. [CrossRef]

10.   Cantero, B.; del Bosque, I.F.S.; Matías, A.; Medina, C. Statistically significant effects of mixed recycled aggregate on the physical-mechanical properties of structural concretes. *Constr. Build. Mater.* **2018**, *185*, 93–101. [CrossRef]

11.   Ghorbania, S.; Sharifi, S.; Ghorbani, S.; Tam, V.W.Y.; de Brito, J.; Kurda, R. Effect of crushed concrete waste's maximum size as partial replacement of natural coarse aggregate on the mechanical and durability properties of concrete. *Resour. Conserv. Recycl.* **2019**, *149*, 664–673. [CrossRef]

12.   Bui, N.K.; Satomi, T.; Takahashi, H. Mechanical properties of concrete containing 100% treated coarse recycled concrete aggregate. *Constr. Build. Mater.* **2018**, *163*, 496–507. [CrossRef]

13.   Liu, C.; Fan, J.C.; Bai, G.L.; Quan, Z.G.; Fu, G.M.; Zhu, C.; Fan, Z.Y. Cyclic load tests and seismic performance of recycled aggregate concrete (RAC) columns. *Constr. Build. Mater.* **2019**, *195*, 682–694. [CrossRef]

14.   Yao, J.W.; Song, Y.P.; Qin, L.K.; Gao, L.X. Mechanical Properties and Failure Criteria of Concrete under Biaxial Tension and Compression. *Adv. Mater. Res.* **2011**, *261*, 252–255. [CrossRef]

15.   Yang, G.; Zhao, Z.Z.; He, X.G. Experimental Study on Strength and Failure Characteristics of Reinforced Concrete Plate under Complex Stresses Conditions. *Appl. Mech. Mater.* **2013**, *357*, 699–704. [CrossRef]

16.   Bai, G.L.; Yin, Y.G.; Liu, C.; Han, Y.Y. Experimental study on bond-slip behavior between section steel and recycled aggregate concrete in steel reinforced recycled concrete structures. *J. Build. Struct.* **2016**, *37*, 139–146. [CrossRef]

17.   Xie, M.; Zheng, S.S. Fractal Analysis on Bond-Slip Behavior between Steel Shape and Concrete in SRC on Axial Pull-Out Test. *Adv. Mater. Res.* **2012**, *382*, 352–355. [CrossRef]

18.   Liang, Y.B.; Li, D.S.; Seyed, M.P.; Kong, Q.Z.; Ing, L.; Song, G.B. Bond-slip detection of concrete-encased composite structure using electro-mechanical impedance technique. *Smart Mater. Struct.* **2016**, *25*, 095003. [CrossRef]

19.   Rana, M.M.; Lee, C.K.; Al-deen, S. A study on the bond stress-slip behavior between engineered cementitious composites and structural steel sections. *Ce/papers* **2017**, *1*, 2247–2256. [CrossRef]

20.   Liu, C.; Fan, Z.Y.; Chen, X.N.; Zhu, C.; Wang, H.Y.; Bai, G.L. Experimental Study on Bond Behavior between Section Steel and RAC under Full Replacement Ratio. *KSCE J. Civ. Eng.* **2019**, *23*, 1159–1170. [CrossRef]

21.   Jendele, L.; Cervenka, J. Finite element modelling of reinforcement with bond. *Comput. Struct.* **2006**, *84*, 1780–1791. [CrossRef]

22.   Zhou, Z.; Dong, Y.; Jiang, P.; Han, D.; Liu, T. Calculation of Pile Side Friction by Multiparameter Statistical Analysis. *Adv. Civ. Eng.* **2019**, *2019*, 12. [CrossRef]

23.   Vilanova, I.; Torres, L.; Baena, M.; Llorens, M. Numerical simulation of bond-slip interface and tension stiffening in GFRP RC tensile elements. *Compos. Struct.* **2016**, *153*, 504–513. [CrossRef]

24.   Chen, Z.P.; Zheng, H.H.; Xue, J.Y.; Su, Y.S. Test and Bond Strength Analysis of Bonded Slip of Shaped Steel Recycled Concrete. *J. Build. Struct.* **2013**, *34*, 130–138.

25.   Liu, C.; Lv, Z.Y.; Bai, G.L.; Yin, Y.G. Experiment study on bond-slip behavior between section steel and RAC in SRRC structures. *Constr. Build. Mater.* **2018**, *175*, 104–114. [CrossRef]

26.   Hwang, J.-Y.; Kwak, H.-G. FE analysis of circular CFT columns considering bond-slip effect: A numerical formulation. *Mech. Sci.* **2018**, *9*, 245–257. [CrossRef]

27.   Al-Rousan, R.Z. Empirical and NLFEA prediction of bond-slip behavior between DSSF concrete and anchored CFRP composites. *Constr. Build. Mater.* **2018**, *169*, 530–542. [CrossRef]

28.   Butler, L.; West, J.S.; Tighe, S.L. The effect of recycled concrete aggregate properties on the bond strength between RCA concrete and steel reinforcement. *Cem. Concr. Res.* **2011**, *41*, 1037–1049. [CrossRef]

29. Liu, C.; Liu, H.W.; Zhu, C.; Bai, G.L. On the mechanism of internal temperature and humidity response of recycled aggregate concrete based on the recycled aggregate porous interface. *Cem. Concr. Compos.* **2019**, *103*, 22–35. [CrossRef]

30. Rao, M.C.; Bhattacharyya, S.K.; Bara, S.V. Influence of field recycled coarse aggregate on properties of concrete. *Mater. Struct.* **2011**, *44*, 205–220.

31. Ferreira, L.; de Brito, J.; Barra, M. Influence of the pre-saturation of recycled coarse concrete aggregates on concrete properties. *Mag. Concr. Res.* **2011**, *63*, 617–627. [CrossRef]

32. Yang, Y. Basic Theory and Application of Steel-Slab Concrete Bond-Slip. Ph.D. Thesis, Xi'an University of Architecture and Technology, Xi'an, China, 2003.

33. Zheng, H.H.; Chen, Z.P.; Su, Y.S. Composition and Strength of Interface Bonding Stress of Steel Recycled Concrete. *J. Tongji Univ.* **2016**, *44*, 1504–1512.

34. Xiao, J.Z.; Lan, Y. Experimental Study on Uniaxial Tensile Properties of Recycled Concrete. *J. Build. Mater.* **2006**, *9*, 154–158. [CrossRef]

35. Yin, Y.G. *Experimental Study on Bond-Slip Performance of Steel-Reinforced Concrete*; Xi'an University of Architecture and Technology: Xi'an, China, 2016.

36. Chen, J.R. *Experimental Study on Bond-Slip Performance of Steel and Recycled Concrete after High Temperature*; Guangxi University: Guangxi, China, 2016.

37. Prepared by the Ministry of Construction of the People's Republic of China. *Code for Design of Concrete Structures*; China Building Industry Press: Beijing, China, 2002.

38. Xiao, J.Z. Experimental Investigation on Complete Stress-Strain Curve of Recycled Concrete under Uniaxial Loading. *J. Tongji Univ. Nat. Sci.* **2007**, *11*, 5–9.

39. Zhou, W.D.; Fu, J.L.; Xiao, J.P.; Liu, B.K. Seismic behavior of joints in recycled concrete frames based on abaqus. *J. Hefei Univ. Technol. Nat. Sci.* **2016**, *39*, 205–210.

40. Yang, Y.; Guo, Z.X.; Xue, J.Y.; Zhao, H.T.; Nie, J.G. Experiment study on bond slip behavior between section steel and concrete in SRC structures. *J. Build. Struct.* **2005**, *26*, 1–9. [CrossRef]

© 2020 by the authors. Licensee MDPI, Basel, Switzerland. This article is an open access article distributed under the terms and conditions of the Creative Commons Attribution (CC BY) license (http://creativecommons.org/licenses/by/4.0/).

Article

Experimental Study on a Prediction Model of the Shrinkage and Creep of Recycled Aggregate Concrete

Zhenyuan Lv [1], Chao Liu [1,*], Chao Zhu [1], Guoliang Bai [1,2] and Hao Qi [1]

[1] College of Civil Engineering, Xi'an University of Architecture and Technology, Xi'an 710055, China; lvzhenyuan@live.xauat.edu.cn (Z.L.); zhuchao@live.xauat.edu.cn (C.Z.); guoliangbai@126.com (G.B.); qihao@xauat.edu.cn (H.Q.)

[2] Key Laboratory of China's Ministry of Education, Xi'an 710055, China

* Correspondence: chaoliu@xauat.edu.cn

Received: 10 September 2019; Accepted: 1 October 2019; Published: 14 October 2019

Featured Application: The effects of the recycled aggregate replacement rate and water–cement ratio on shrinkage and creep properties were analyzed. The degree of shrinkage and creep of ordinary concrete and recycled aggregate concrete at different ages were compared. A coefficient of increase and a coefficient of contraction increase of the attached mortar were proposed. Based on the old attached mortar, a shrinkage and creep model of recycled aggregate concrete was established.

Abstract: The significant difference between recycled aggregate and natural aggregate is the content of the attached mortar layer. With the increase of the replacement rate of recycled aggregate, the shrinkage and creep of recycled aggregate concrete is significantly increased. In this paper, 180-day shrinkage and creep tests of recycled aggregate concrete with different water–cement ratios were designed in order to analyze the effect of the substitution rate and water–cement ratio on shrinkage and creep properties. The results show that the shrinkage strain of recycled aggregate concrete with a substitution rate of 50% and 100% at 180 days is 26% and 48% higher than that of ordinary concrete, respectively, and the growth of group II is 22% and 47%, respectively. When the load was 180 days old, the creep coefficient of recycled aggregate concrete with a substitution rate of 50% and 100% in group I increased by 19.6% and 39.6%, respectively compared with ordinary concrete, and group II increased by 23.6% and 44.3%, respectively. Based on the difference of adhering mortar content, the creeping increase coefficient and shrinkage increase coefficient of the attached mortar were proposed, and a shrinkage and creep model of recycled aggregate concrete was established. When compared with the experimental results, the model calculation results met the accuracy requirements.

Keywords: recycled aggregate concrete; shrinkage and creep; attached mortar; prediction model

---

1. Introduction

The recycling of construction waste (RCW) refers to the crushing of construction waste to obtain different types of products and reuse them as resources. An important product of RCW is recycled aggregate, which itself is attached to old mortar that is difficult to peel due to the limitation of crushing technology. Therefore, it can be approximated that recycled aggregate is a two-phase material composed of the old mortar and the natural aggregate wrapped therewith. Recycled aggregate concrete (RAC) refers to recycled aggregate, which is made by crushing and classifying waste concrete and mixing according to a certain proportion, and partially or completely replacing the natural aggregate to make new concrete referred to as RAC [1]. RAC not only digests waste concrete and solves the problem of a serious shortage of natural resources, but also meets the requirements of existing specifications under reasonable design. It is a green building material worthy of promotion. At present, domestic and foreign countries have given eager attention and long-term exploration of RAC, especially towards the shrinkage and creep of the concrete, which can cause structural deterioration and safety problems. Based on a number of research reports on concrete shrinkage and creep [2–4], the effects of concrete shrinkage and creep mainly include the increase of beam deflection and the reduction of structural bearing capacity. In addition, the shrinkage and creep of the concrete can also cause prestressing loss of the prestressed members and secondary internal forces that create the structure. In the local area of the concrete, due to the shrinkage of internal stress, it is easy for problems such as cracks on the outer surface of the member to be created. In view of the fact that RAC has properties close to that of ordinary concrete [5–12], the structural adverse effects caused by shrinkage and creep are similar or even more serious than in natural concrete. Therefore, the research on the shrinkage and creep of RAC has an irreplaceable significance.

Many scholars have conducted relevant research on the importance of the shrinkage and creep of RAC. Domingo-Cabo et al. [13] studied the shrinkage and creep tests of RAC with different substitution rates. The results show that with the increase of the replacement rate of recycled aggregates, the shrinkage and creep deformation of RAC increases. However, there is a lack of in-depth exploration of the impact of factors. Geng et al. [14] conducted experimental research on the creep behavior of RAC with different water–cement ratios of base concrete. The results show that the creep properties of recycled base concrete with a low water–cement ratio are significantly affected by recycled coarse aggregate (RCA), and the establishment of a prediction model for the creep of recycled base concrete needs to consider the influence of the water–cement ratio. However, the effects of different recycled aggregate-attached mortars on shrinkage and creep have not been deeply considered. Adam et al. [15] studied the shrinkage and creep of RAC based on three factors: curing condition, loading age, and axial stress level. The experimental results show that the corresponding average ratios of the shrinkage strain of RAC of a 50% and 100% substitution rate are 1.21 and 1.71, compared to ordinary concrete. Guo et al. [16] devoted himself to the study of the effect of old adhesive mortar on the creep of RAC. Based on the experimental data, it is proposed that the RAC with a 100% substitution rate develops more rapidly and the amplitude is 1.6 times larger than that of traditional concrete. Miguel et al. [17] explored the effect of recycled aggregates from Portuguese construction and demolition waste on the shrinkage and creep properties of concrete. The results showed that the creep coefficient of the 12 recycled aggregates tested at 91 days was 0.22 to 1.63. This proves that the shrinkage prediction model of ordinary concrete is not suitable for RAC, so research on a prediction model of the shrinkage and creep of RAC has become the focus of solving the problem of the shrinkage and creep of RAC.

Fathifazl et al. [18] pointed out that the prediction model of RAC creep should consider the influence of mortar attached to the aggregate surface. Based on the MC90 model, the predicted values of the model agree well with the experimental values. Brito et al. [19] studied the Gómez-Soberón test data, and the concrete creep coefficient was corrected considering the difference in apparent density or water absorption between recycled aggregate and natural aggregate. Based on the European standard EC2 model, Brito (D) and Brito (R) obtained two kinds of RAC creep prediction models. The prediction accuracy of the two models is good. Tošic et al. [20] explored the effect of recycled aggregates on the creep of concrete. It was pointed out that when using the creep prediction model of fib Model Code 2010 to predict the creep coefficient of RAC, the creep coefficient of RAC is underestimated relative to the performance of the model on the accompanying natural aggregate concrete (NAC). At the same time, compared with NAC, RAC shows a larger creep coefficient, and the average increase in the coefficient of creep is 39% for RAC at full replacement rate. Silva et al. [21] studied the correction coefficient of the concrete creep of different recycled aggregates and a prediction model suitable for recycled aggregate concrete. Compared with conventional concrete, the use of recycled aggregate to absorb part of the mixed water and cement slurry for coating produces a stronger ITZ (Interface transition zone), which can reduce the creep strain by up to 23%. Liu et al. [22] explored the calculation method of the long-term deformation of recycled concrete beams based on the creep adjustment coefficient. At the same time, the adjustment coefficient of the mortar creep bond was proposed. Three typical ordinary concrete shrinkage and creep prediction models were modified and used for long-term deformation calculation of recycled concrete beams. Luo et al. [23] proposed a model for the shrinkage and creep of RAC that considered the two factors of the regenerated aggregate grade and substitution rate, and the regression coefficient was used to obtain the influence coefficient expression. This can quantitatively calculate the influence coefficient of the shrinkage and creep of recycled aggregates with different quality and different substitution rates. Luo et al. [24] studied the influence coefficient of aggregate pretreatment on the creep behavior of RAC. Xiao et al. [25] corrected the q2 and q4 parameters in the B3 model based on the test results. Considering the influence of the replacement rate of the RCA, the calculated results of the modified model were in good agreement with the experimental results.

However, the existing method of proposing the correction coefficient by a certain influencing factor fails to establish the prediction model of the shrinkage and creep of RAC. In particular, for the effect of recycled aggregate-attached mortar on concrete shrinkage and creep, there is a lack of predictive models with universal significance. In this paper, the key difference between recycled aggregate and natural aggregate-attachment mortar is taken as the entry point, and the coefficient of increase and the coefficient of contraction increase of the attached mortar are proposed. Based on the two increasing factors, the shrinkage and creep model of RAC is established, and the applicability of the model is checked.

2. Shrinkage and Creep Test of Recycled Aggregate Concrete

2.1. Materials

The test materials were selected from PC32.5R ordinary Portland cement, natural coarse aggregate, RCA, natural fine aggregate, ordinary fine sand, and urban tap water. The RCA was provided by Shaanxi Science and Technology Environmental Protection Co., Ltd. (Xi'an, China), and its physical properties are shown in Table 1.

Table 1. Physical properties of recycled coarse aggregate (RCA) and natural coarse aggregate (NCA).

Aggregate Type	Apparent Density/kg·m⁻³	Crushing Index/%	Moisture Content/%	Water Absorption Rate/%	Initial Stone/%	Secondary Aggregate/%	Mortar Block/%	Impurity/%
Recycled coarse aggregate	2458	17.0	1.33	3.83	29.5	51.2	16.0	3.3
Natural coarse aggregate	2658	10.6	4.21	0.69	98.7	—		1.3

2.2. Mix Proportion Design

According to JGJ55-2011 [26] "Ordinary concrete mix design rules", the mixing ratio of natural aggregate concrete (NAC) was calculated. The mixing ratio of RAC was replaced by equal mass. The natural coarse aggregate in ordinary concrete was replaced by RCA. Considering that the water absorption of the RCA is higher than that of the natural coarse aggregate, a certain amount of additional water was added, which was 3.83% of the mass of the RCA. The concrete slump of each component during mixing is about 90 mm. The proportions of each group and the compressive strength of the concrete cubes are shown in Table 2.

Table 2. Mix ratio of recycled aggregate concrete (RAC) and natural aggregate concrete (NAC).

Group Number	r/%	Effective Water–Cement Ratio	Material Consumption/kg·m^{-3}						Compressive Strength/N/mm^2
			Cement	Water	Natural Coarse Aggregate	Recycled Coarse Aggregate	Fine Aggregate	Additional Water	
NAC-I	0	0.527	370	195	1185	-	660	-	33.3
RAC-50-I	50	0.527	370	195	592.5	592.5	660	22.69	31.6
RAC-100-I	100	0.527	370	195	-	1185	660	45.38	32.4
NAC-II	0	0.40	500	200	1086	-	611	-	35.1
RAC-50-II	50	0.40	500	200	543	543	611	20.8	32.3
RAC-100-II	100	0.40	500	200	-	1086	611	41.6	30.9

Note: Group I was the water–cement ratio (w/c) = 0.527 sample; Group II was the water–cement ratio = 0.4 sample. The additional water was calculated to take 3.83% of the weight of the recycled coarse aggregate under consideration of the water absorption rate of the recycled coarse aggregate.

2.3. Specimen Design

The creep test of RAC was carried out in accordance with "Test Method for Long-term Performance and Durability of Ordinary Concrete" [27]. The creep specimen size was 100 mm × 100 mm × 400 mm, the shrinkage specimen size was 100 mm × 100 mm × 515 mm, cubic specimen size was 150 mm × 150 mm × 150 mm, and prismatic specimen size was 150 mm × 150 mm × 300 mm. The specific number of specimens and uses is shown in Table 3.

Table 3. Number and use of specimens.

Test Piece Number	Specimen Number				Specimen Use
	Cube	Prism	Shrinkage Specimen	Creep Specimen	
NAC-I	3	6	3	2	
RAC-50-I	3	6	3	2	A Cube: Cube used to determine the compressive strength.
RAC-100-I	3	6	3	2	B Prisms: Three were used to determine the
NAC-II	3	6	3	2	ultimate bearing capacity and three were
RAC-50-II	3	6	3	2	used to determine the elastic modulus.
RAC-100-II	3	6	3	2	

2.4. Test Loading Process

The creep testing of RAC was carried out using a spring-loaded compression creeper that was loaded with a jack. The load was controlled by a pressure sensor and a digital electronic displacement meter, and the constant value of the load was maintained by the spring reaction force. The creep loading device and shrink test device are shown in Figure 1.

(1) The creep specimens were maintained for 28 days. The prism compressive strength of the specimens under the same conditions was tested before loading. The electronic displacement meter was checked for zero and the initial reading was recorded.

(2) After the completion of the alignment, it started loading in time and the creep stress was taken as 60% of the measured prismatic compressive strength.

(3) The deformation values of the test piece were read at 1 day, 3 days, 7 days, 14 days, 28 days, 45 days, 60 days, 90 days, 120 days, 150 days, and 180 days after loading, and the shrinkage value of the shrinkage test piece in the same environment was recorded.

(4) The load was checked regularly after loading. If the load changed by more than 2%, the correct load was applied, and the nut on the screw was tightened to make up for it.

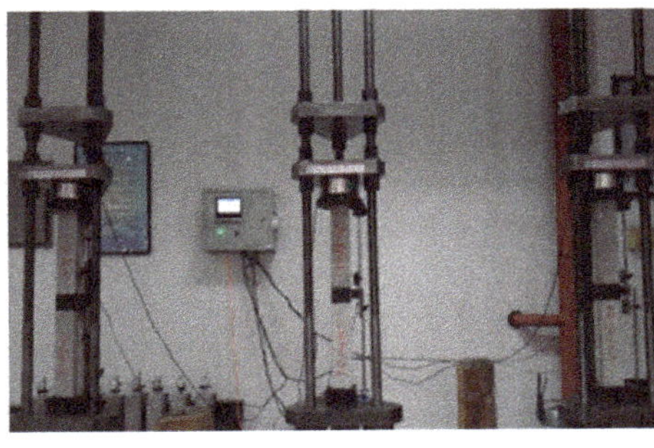

(a) Creep loading device   (b) Shrinkage test device

Figure 1. Site loading schematic.

3. Test Results and Analysis

3.1. Analysis of Shrinkage Test Results

The shrinkage of RAC with different substitution rates at 180 days is shown in Figure 2. The shrinkage of RAC is similar to that of normal concrete. At 180 days, the shrinkage of RAC with a 50% and 100% substitution rate in group I increased by 26% and 48%, respectively, compared with that of ordinary concrete, and the shrinkage of RAC with a 50% and 100% substitution rate in group II was increased by 22% % and 47%, respectively, compared with that of ordinary concrete. As the replacement rate of RCA increased, the shrinkage of RAC also increased. In the early stage of shrinkage, the shrinkage of RAC increased faster and shrank faster, and then the rate of shrinkage decreased and slowed down. The shrinkage tended to be gentle until 120 days, and about 95% of the total shrinkage was achieved. It was considered that the shrinkage of RAC tended to be stable at 180 days.

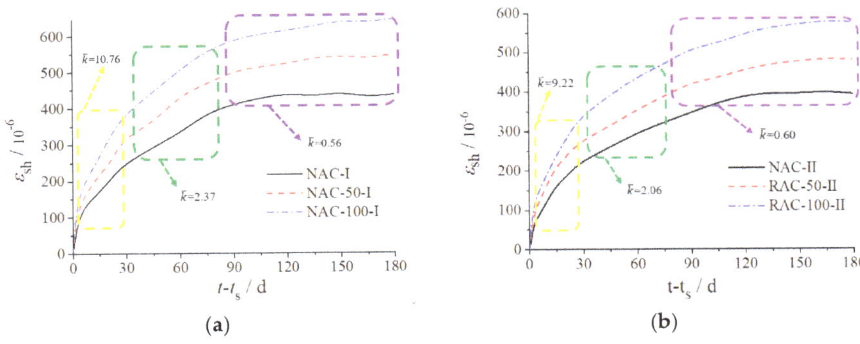

Figure 2. Shrinkage deformation curve of RAC with different water–cement ratios. (**a**) *w/c* = 0.527; (**b**) *w/c* = 0.4. Note: ε_{sh} indicates shrinkage strain and is dimensionless; $\bar{k}$ indicates the slope and is dimensionless; t_s indicates the age of the concrete at the start of drying; t indicates the age of the concrete.

In the early stage of shrinkage, the shrinkage deformation of normal concrete with *w/c* = 0.4 was close to that of normal concrete with *w/c* = 0.527, and then the deviation between them gradually increased. Normal concrete shrinkage with *w/c* = 0.527 at 180 days was 11% higher than normal concrete shrinkage with *w/c* = 0.4. At 180 days, the shrinkage deformation of RAC with *w/c* = 0.527 increased by 14% and 12%, respectively, compared with that of RAC with a substitution rate of 50% and 100% with *w/c* = 0.4. This shows that the water–cement ratio is not the main factor affecting the shrinkage of RAC.

As the age increased, the shrinkage of RAC was not affected by the water–cement ratio. In addition, both groups of RAC and ordinary concrete showed an obvious shrinkage development trend in the first 30 days: The average shrinkage slope reached 10.76 and 9.22, respectively; the shrinkage development trend of each group of samples slowed down and contracted from 30 days to 86 days, and the average slope decreased to 2.37 and 2.06, respectively; the shrinkage trend of each group tended to be stable from 90 days to 180 days, and the average shrinkage slopes approached 0.56 and 0.60, respectively. Comparing the changes in the average shrinkage slopes of the two water–cement ratios, it was found that the water–cement ratio had a more obvious influence on the trend of early shrinkage. The average shrinkage slope of the $w/c = 0.527$ sample increased by 16.7% in the first stage ($\leq$30 days) compared to the $w/c = 0.4$ sample; the second stage of the increase in value accounted for 17.7% of the first stage; the third stage of the average slope slightly decreased to account for 2.3% of the first stage.

Similarly to ordinary concrete, the main cause of the shrinkage of RAC is cement mortar hardening, internal moisture loss, and deformation caused by volume reduction. However, the shrinkage of RAC is higher than that of ordinary concrete, and as the substitution rate increases, the amount of shrinkage also increases. On the one hand, this is because the micro-cracks inside the RCA cause a decrease in the elastic modulus and strength, resulting in a reduced ability to restrain the shrinkage, a smaller volume loss, and a larger shrinkage deformation. On the other hand, compared with ordinary concrete, due to the presence of the old cement mortar, which is not easily peeled off on the surface of the RCA, the total amount of mortar of the RAC is larger under the same mix ratio, and the shrinkage is mainly caused by the hardening of the cement mortar. Therefore, the shrinkage of RAC in the same environment is stronger than that of ordinary concrete.

3.2. Analysis of Creep Test Results

The creep curve of RAC at 180 days is shown in Figure 3. The creep of RAC is similar to that of ordinary concrete. At 180 days, the creep coefficient of RAC with a 50% and 100% substitution rate in group I increased by 19.6% and 39.6%, respectively, compared with that of ordinary concrete. The creep coefficient of RAC with a 50% and 100% substitution rate in group II was 23.6% and 44.3% higher than that of ordinary concrete, respectively. At the initial stage of loading, the creep of RAC with a 50% and 100% substitution rate increased faster than that of ordinary concrete, and the creep coefficient was always higher than ordinary concrete. The creep growth rate decreased with the passage of time, but the decrease rate of the creep growth rate of ordinary concrete that was close to the creep coefficient of RAC was small. At 120 days, the creep development of all three creeps tended to be flat, and about 90% of the total creep variable was completed, which roughly shows that the creep of the RAC tended to be stable.

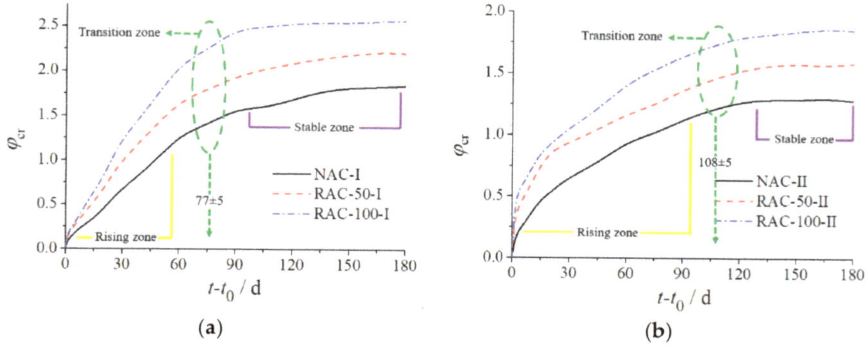

Figure 3. Creep coefficient curve of RAC. (**a**) $w/c = 0.527$; (**b**) $w/c = 0.4$. Note: ε_{sh} indicates shrinkage strain and is dimensionless; $\bar{k}$ indicates the slope and is dimensionless; t_0 indicates the loading age of concrete; t indicates the age of the concrete.

At the initial stage of loading, the creep of normal concrete with $w/c = 0.4$ was faster than that of normal concrete with $w/c = 0.527$, and the creep coefficient was higher than the latter. After that, the creep growth rate of normal concrete with $w/c = 0.4$ decreased and that of ordinary concrete with $w/c = 0.527$ increased rapidly. When loading to 180 days, the creep coefficient of normal concrete with $w/c = 0.527$ increased by 43.8% compared with that of RAC with $w/c = 0.4$. For the RAC with $w/c = 0.4$ and a substitution rate of 50% and 100%, the creep development trend was the same. At 180 days, the creep coefficient of RAC with $w/c = 0.527$ increased by 40% and 38.7%, respectively, compared with that of RAC with a substitution rate of 50% and 100% with $w/c = 0.4$. This shows that the water–cement ratio is an important factor affecting the creep of RAC. The high water–cement ratio and the large amount of recycled mortar have a significant effect on creep.

4. Creep and Shrinkage Model of Recycled Aggregate Concrete

At present, there are many models of shrinkage and creep of ordinary concrete at home and abroad. In this paper, five typical models are selected, which are the CEB-FIP 1990 model [28], ACI209R 1992 model [29], GL2000 model [30], B3 model [31], and GB50010 model [32]. Based on the five models, the recycled aggregate attached mortar is taken as the research object, the increasing coefficient of the attached mortar is put forward, and the shrinkage creep model of the RAC is established.

4.1. Creep Increasing Coefficient of Attached Mortar

Fathifazl et al. [18] considered that natural aggregate was replaced by mass or volume percentage when RAC was prepared by traditional methods. Compared with ordinary concrete, a certain amount of cement mortar is wrapped on the surface of the recycled aggregate. The content of natural aggregate in the RAC prepared by the conventional method is decreased and the content of the total mortar is increased. The elastic modulus of the component is decreased, and the shrinkage and the amount of creep are increased. It is assumed that the RCA is a two-phase material composed of residual cement mortar and natural aggregate, and the mortar in the RAC consists of residual cement mortar and new cement mortar on the aggregate surface. Based on the creep test results of RAC, the residual cement mortar content of RCA was taken as the research object to establish the creep prediction model of RAC.

$$C_{AM} = \frac{C_{RAC}}{C_{NAC}} = \left(\frac{V_{NM}^{RAC} + V_{RM}^{RAC}}{1 - V_{RCA}^{RAC}} \right)^{1.33}, \tag{1}$$

$$V_{NM}^{RAC} = 1 - V_{RCA}^{RAC}. \tag{2}$$

Simultaneous equations:

$$C_{AM} = \left(\frac{V_{NM}^{RAC} + V_{AM}^{RAC}}{1 - V_{NCA}^{NAC} - V_{RCA}^{NAC}} \right)^{1.33}. \tag{3}$$

According to the composition of RCA, the following equation holds:

$$v_{RCA} = v_{AM} + v_{OVA}, \quad m_{RCA} = m_{AM} + m_{OVA}, \tag{4}$$

$$m_{RCA} = v_{RCA} \times \rho_{RCA}, \quad m_{AM} = v_{AM} \times \rho_{AM}, \tag{5}$$

$$m_{OVA} = v_{OVA} \times \rho_{OVA}. \tag{6}$$

In the equation, V is volume (m^3); m is quality (kg); ρ is apparent density (kg/m^3).

The parameter M_{AM} is the mass ratio of the attached mortar and the RCA. $M_{AM} = m_{AM}/m_{RCA}$, and the following equation is derived from the above equation:

$$\rho_{AM} = \frac{m_{AM}}{v_{AM}} = \frac{v_{RCA} \times \rho_{RCA} \times M_{AM}}{v_{RCA} - \frac{(1-M_{AM})v_{RCA} \times \rho_{RCA}}{\rho_{OVA}}} = \frac{M_{AM}}{\frac{1}{\rho_{OVA}} - \frac{1-M_{AM}}{\rho_{OVA}}}, \tag{7}$$

and $v_{AM} = v_{RCA} \times M_{AM} \times \frac{\rho_{RCA}}{\rho_{AM}}$.

Simultaneous equations:

$$v_{AM} = v_{RCA} \times [1 - (1 - M_{AM})\frac{\rho_{RCA}}{\rho_{OVA}}], \tag{8}$$

$$V_{AM} = V_{RCA} \times [1 - (1 - M_{AM})\frac{\rho_{RCA}}{\rho_{OVA}}]. \tag{9}$$

Generally, the replacement rate of RCA is defined as follows:

$$r = \frac{m_{RCA}^{RAC}}{m_{RCA}^{RAC} + m_{NCA}^{RAC}}. \tag{10}$$

The volume ratio of natural coarse aggregate and RCA in RAC is R.

$$R = \frac{v_{NCA}^{RAC}}{v_{RCA}^{RAC}} = \frac{V_{NCA}^{RAC}}{V_{RCA}^{RAC}} = \frac{(m_{RCA}^{RAC} + m_{NCA}^{RAC}) \times (1-r)/\rho_{NCA}}{(m_{RCA}^{RAC} + m_{NCA}^{RAC}) \times r/\rho_{RCA}} = \frac{1-r}{r} \times \frac{\rho_{RCA}}{\rho_{NCA}} \tag{11}$$

Equation (3) can be rewritten as

$$C_{AM} = \left(\frac{1 - V_{RCA} \times [R + (1 - M_{AM})\frac{\rho_{RCA}}{\rho_{OVA}}]}{1 - V_{RCA}^{RAC} \times (1 + R)}\right)^{1.33}. \tag{12}$$

In order to simplify the expression, the approximate $\rho_{RCA} = \rho_{OVA}$ is considered, and the relationship between C_{AM} and the RCA replacement rate r is as follows:

$$C_{AM} = \left(\frac{1 - V_{RCA}^{RAC} \times [\frac{1-r}{r} \times \frac{\rho_{RCA}}{\rho_{NCA}} + (1 - M_{AM})]}{1 - V_{RCA}^{RAC} \times (1 + \frac{1-r}{r} \times \frac{\rho_{RCA}}{\rho_{NCA}})}\right)^{1.33}. \tag{13}$$

When r = 100%, it is all RAC:

$$C_{AM} = \left(\frac{1 - V_{RCA}^{RAC}(1 - M_{AM})}{1 - V_{RCA}^{RAC}}\right)^{1.33}. \tag{14}$$

4.2. Shrinkage Increasing Coefficient of Attached Mortar

Concrete is constrained by aggregates, so its dry shrinkage is lower than that of pure grout [33]. The shrinkage of concrete is mainly caused by the hydration of cement mortar. The size of a few coarse aggregates is unstable. However, most of the coarse aggregates have the same size, a large elastic modulus, high crush index, and good restraint on concrete shrinkage. Therefore, the shrinkage of concrete depends on the amount and rigidity of the coarse aggregate in concrete. Considering the combined effect of aggregate content and rigidity, the prediction equation of the shrinkage strain of ordinary concrete is as follows:

$$S_{NAC} = S_{TP}^{NAC}(V_{TP}^{NAC})^{\alpha}. \tag{15}$$

In the equation, S_{NAC}—the shrinkage strain of ordinary concrete; S_{TP}—the shrinkage strain of cement slurry in common concrete under the same conditions; V_{TP}^{NAC}—the volume of the total cement slurry in the ordinary concrete; α—empirical coefficient. The change range is 1.2–1.7 and the average is 1.45.

Because of the two-phase nature of RCA, the total mortar content in RAC is composed of attached mortar and fresh mortar. The shrinkage strain of RAC can be expressed as the following equation:

$$S_{RAC} = S_{TP}^{RAC}(V_{TP}^{RAC})^{\alpha}. \tag{16}$$

It is assumed that the shrinkage characteristics of cement mortar in RAC are the same as those of ordinary concrete, which simultaneously gives the following equation:

$$\frac{S_{RAC}}{S_{NAC}} = \left(\frac{V_{TP}^{RAC}}{V_{TP}^{NAC}}\right)^{\alpha}. \tag{17}$$

The definition of SAM is the shrinkage increase coefficient of adhesive mortar, $S_{AM} = (V_{TP}^{RAC}/V_{TP}^{NAC})^{\alpha}$. Since the shrinkage of RAC is related to the volume of the total mortar, the volume V_{TP}^{RAC} of the total mortar in RAC can be determined according to Equation (9). Then there is the following equation:

$$S_{AM} = \left(\frac{1 - V_{RCA}^{RAC} \times \left[\frac{1-r}{r} \times \frac{\rho_{RCA}}{\rho_{NCA}} + (1 - M_{AM})\right]}{1 - V_{RCA}^{RAC} \times \left(1 + \frac{1-r}{r} \times \frac{\rho_{RCA}}{\rho_{NCA}}\right)}\right)^{1.45}. \tag{18}$$

When $r = 100\%$, it is all RAC:

$$S_{AM} = \left(\frac{1 - V_{RCA}^{RAC}(1 - M_{AM})}{1 - V_{RCA}^{RAC}}\right)^{1.45}. \tag{19}$$

5. Conclusions

(1) The development law of the shrinkage and creep of RAC is similar to that of ordinary concrete. At 180 days, compared with ordinary concrete, the shrinkage of group I RAC with a substitution rate of 50% and 100% was increased by 26% and 48%, respectively, and the group II RAC was increased by 22% and 47%, respectively. When the load was 180 days old, compared with ordinary concrete, the creep rate of group I RAC with a substitution rate of 50% and 100% was increased by 19.6% and 39.6%, respectively, and group II was increased by 23.6% and 44.3%, respectively. With the increase of the replacement rate of recycled aggregate, the shrinkage and creep of RAC increased significantly, which indicates that the substitution rate is an important factor affecting the shrinkage and creep of RAC, and the water–cement ratio has more significant effects on the creep of RAC.

(2) The effect of the substitution rate on the shrinkage and creep of RAC is caused by the adhesion of mortar. As the substitution rate increases, the porosity of the attached mortar increases the degree of concrete shrinkage. The increase of the water–cement ratio leads to an increase in the proportion of the attached mortar in the RAC. More adherence of the mortar pores weakens the restraint performance of the RAC, which in turn reduces the resistance of the RAC to creep.

(3) The key difference between recycled aggregate and natural aggregate is the difference in the content of the attached mortar. In this paper, the contraction point was used to calculate the shrinkage strain and creep coefficient of the RAC shrinkage model, and the attached mortar increase coefficient method was also used. The calculated value of the model takes into account the effect of the recycled mortar itself on the shrinkage and creep of the recycled concrete. The difference between the calculated value and the experimental value is small and can meet the accuracy requirements well.

(4) Based on the difference of the attached mortar, effective shrinkage and creep of RAC can be established, and the shrinkage and creep of RAC will show a different development trend from ordinary concrete as the age increases. At the same time, according to the same research ideas, a complete prediction model of the shrinkage and creep of RAC with a full-service life cycle can

be established, and the calculation method of the long-term deformation of RAC with significant influence on shrinkage and creep can be carried out.

Author Contributions: Z.L. and C.L. were responsible for the writing of this article. C.Z. and G.B. were responsible for the evaluation and comments of the paper content, and H.Q. was responsible for the organization of the article data.

Funding: This research was funded by [the Natural Science Foundation of China] grant number [51878546], [the Shaanxi Science and Technology Innovation Base] grant number [2017KTPT-19], [the Shaanxi Province Innovative Talent Promotion Plan Project] grant number [2018KJXX-056], the Shaanxi Provincial Key Research and Development Program] grant number [2018ZDCXL-SF-03-03-02].

Acknowledgments: The support from the following fund projects is highly acknowledged: (1) the Natural Science Foundation of China (51878546); (2) the Shaanxi Science and Technology Innovation Base (2017KTPT-19); (3) the Shaanxi Province Innovative Talent Promotion Plan Project (2018KJXX-056); (4) the Shaanxi Provincial Key Research and Development Program (2018ZDCXL-SF-03-03-02).

Conflicts of Interest: All authors declare that this article has no conflict of interest, and the data (including figures, tables) and other relevant information used in the article are approved by all authors. The correspondent authors exercise power over the article on behalf of all authors.

References

1. Xiao, J.Z. *Recycled Aggregate Concrete*; China Building Industry Press: Beijing, China, 2008.
2. Huang, G.X.; Hui, R.Y.; Wang, X.J. *Concrete Creep and Shrinkage*; China Electric Power Press: Beijing, China, 2012.
3. Asamoto, S.; Ohtsuka, A.; Kuwahara, Y.; Miura, C. Study on effects of solar radiation and rain on shrinkage, shrinkage cracking and creep of concrete. *Cem. Concr. Res.* **2011**, *41*, 590–601. [CrossRef]
4. Ma, J.X.; Dehn, F. Shrinkage and creep behavior of an alkali-activated slag concrete. *Struct. Concr.* **2017**, *18*, 801–810. [CrossRef]
5. Xiao, J.Z.; Li, W.; Fan, Y.H.; Huang, X. An overview of study on recycled aggregate concrete in China (1996–2011). *Constr. Build. Mater.* **2012**, *31*, 364–383. [CrossRef]
6. Ozbakkaloglu, T.; Gholampour, A.; Xie, T.Y. Mechanical and Durability Properties of Recycled Aggregate Concrete: Effect of Recycled Aggregate Properties and Content. *J. Mater. Civ. Eng.* **2018**, *30*, 2. [CrossRef]
7. Kou, S.C.; Poon, S.C. Enhancing the durability properties of concrete prepared with coarse recycled aggregate. *Constr. Build. Mater.* **2012**, *35*, 69–76. [CrossRef]
8. Zhou, C.H.; Chen, Z.P. Mechanical properties of recycled concrete made with different types of coarse aggregate. *Constr. Build. Mater.* **2017**, *134*, 497–506. [CrossRef]
9. Pandurangan, K.; Dayanithy, A.; Prakash, S.O. Influence of treatment methods on the bond strength of recycled aggregate concrete. *Constr. Build. Mater.* **2016**, *120*, 212–221. [CrossRef]
10. Seara-Paz, S.; Gonzalez-Fonteboa, B.; Martinez-Abella, F.; Gonzalez-Taboada, I. Time-dependent behaviour of structural concrete made with recycled coarse aggregates Creep and shrinkage. *Constr. Build. Mater.* **2016**, *122*, 95–109. [CrossRef]
11. He, S.Y.; Lai, J.X.; Wang, L.X.; Wang, K. A literature review on properties and applications of grouts for shield tunnel. *Constr. Build. Mater.* **2020**, *231*, 468–482.
12. Li, P.L.; Lu, Y.Q.; Lai, J.X.; Liu, H.Q. A comparative study of protective schemes for shield tunneling adjacent to pile groups. *Adv. Civ. Eng.* **2020**, *12*, 1874137.
13. Domingo-Cabo, A.; Lázaro, C.; López-Gayarre, F. Creep and shrinkage of recycled aggregate concrete. *Constr. Build. Mater.* **2009**, *23*, 2545–2553. [CrossRef]
14. Geng, Y.; Wang, Y.; Chen, J. Creep behaviour of concrete using recycled coarse aggregates obtained from source concrete with different strengths. *Constr. Build. Mater.* **2016**, *128*, 199–213. [CrossRef]
15. Adam, M.K.; Yahya, C.K. Creep and Shrinkage of Normal-Strength Concrete with recycled aggregate concrete Aggregates. *ACI Mater. J.* **2015**, *3*, 112. [CrossRef]
16. Guo, M.H.; Frédéric, G.; Ahmed, L. Numerical method to model the creep of recycled aggregate concrete by considering the old attached mortar. *Cem. Concr. Res.* **2019**, *118*, 14–24. [CrossRef]
17. Miguel, B.; Jorge, P.; Jorge, B.; Luís, E. Shrinkage and creep performance of concrete with recycled aggregates from CDW plants. *Mag. Concr. Res.* **2017**, *3*, 20. [CrossRef]

18. Fathifazl, G.; Razaqpur, A.G.; Isgor, O.B. Creep and drying shrinkage characteristics of concrete produced with coarse recycled concrete aggregate. *Cem. Concr. Compos.* **2011**, *33*, 1026–1037. [CrossRef]

19. De Brito, J.; Robles, R. Recycled aggregate concrete methodology for estimating its long-term properties. *Indian J. Eng. Mater. Sci.* **2010**, *17*, 449–462. [CrossRef]

20. Tošić, N.; de la Fuente, A.; Marinković, S. Creep of recycled aggregate concrete: Experimental database and creep prediction model according to the fib Model Code 2010. *Constr. Build. Mater.* **2019**, *195*, 590–599. [CrossRef]

21. Silva, R.; De Brito, J.; Dhir, R. Comparative analysis of existing prediction models on the creep behaviour of recycled aggregate concrete. *Eng. Struct.* **2015**, *100*, 31–42. [CrossRef]

22. Liu, C.; Lv, Z.Y.; Zhu, C.; Bai, G.L.; Zhang, Y. Study on Calculation Method of Long Term Deformation of RAC Beam based on Creep Adjustment Coefficient. *KSCE J. Civ. Eng.* **2019**, *23*, 260–267. [CrossRef]

23. Luo, J.L.; Xu, Z.S.; Xiong, W. Comparative Experimental Study on Shrinkage and Creep of Recycled High-performance Concrete. *Ind. Constr.* **2014**, *44*, 79–83. (In Chinese) [CrossRef]

24. Luo, S.R.; Zheng, X.; Huang, H.S. Experimental study on pretreatment of recycled coarse aggregate and creep of recycled aggregate concrete. *J. Build. Mater.* **2016**, *19*, 242–247. (In Chinese) [CrossRef]

25. Xiao, J.Z.; Xu, X.D.; Fan, Y.H. The Shrinkage and Creep Test of recycled aggregate concrete and Prediction of Creep Neural Network. *J. Build. Mater.* **2013**, *16*, 752–757. (In Chinese) [CrossRef]

26. Ministry of Housing and Urban-Rural Construction of the People's Republic of China. *Design Rules for Mix Proportion of Ordinary Concrete*; Standard No. JGJ55-2011; China Building Industry Press: Beijing, China, 2011.

27. *GB/T50082-2009. Test Method Standard for Long-Term Performance and Durability of Ordinary Concrete*; China Building Industry Press: Beijing, China, 2009.

28. *CEB-FIP Model Code for Concrete Structures*; Committee Euro-international du Beton/Federation International de la Precon-straine: Paris, France, 1990.

29. ACI Committee 209. *Prediction of Creep Shrinkage and Temperature Effects in Concrete Structures (209R-92)*; America Concrete Institute: Farmington Hills, MI, USA, 1992.

30. Gardner, N.J.; Lockman, M.J. Design Provisions for Drying shrinkage and creep of normal-strength concrete. *ACI Mater. J.* **2001**, *98*, 159–167. [CrossRef]

31. Bažant, Z.P.; Baweja, S. Creep and shrinkage prediction model for analysis and design of concrete structures-model B3. *Mater. Struc.* **2000**, *29*, 126. [CrossRef]

32. *GB50010-2010. Code for Design of Concrete Structures*; China Building Industry Press: Beijing, China, 2010.

33. Mindess, S.; Young, J.F. *Concrete*; Prentice-Hall Incorporation: Upper Saddle River, NJ, USA, 1981.

 © 2019 by the authors. Licensee MDPI, Basel, Switzerland. This article is an open access article distributed under the terms and conditions of the Creative Commons Attribution (CC BY) license (http://creativecommons.org/licenses/by/4.0/).

MDPI

St. Alban-Anlage 66

4052 Basel

Switzerland

Tel. +41 61 683 77 34

Fax +41 61 302 89 18

www.mdpi.com

Applied Sciences Editorial Office

E-mail: applsci@mdpi.com

www.mdpi.com/journal/applsci

www.ingramcontent.com/pod-product-compliance
Lightning Source LLC
LaVergne TN
LVHW070650170726
843515LV00004B/376